Flying Dinosaurs

Flying Dinosaurs

How Fearsome Reptiles
Became Birds

John Pickrell

Columbia University Press
New York

Columbia University Press
Publishers Since 1893
New York   Chichester, West Sussex
cup.columbia.edu
Copyright © 2014 John Pickrell
All rights reserved

First published in Australia by NewSouth, an imprint of the University of New South Wales Press, Ltd.

ISBN 978-0-231-17178-6 (cloth : alk. paper)
ISBN 978-0-231-53878-7 (e-book)
Library of Congress Control Number : 2014938400

Book design: Josephine Pajor-Markus
Cover design: Xou Creative
Front cover images: Feathered dinosaur Guanlong wucaii faces off against its modern relative and 'flying dinosaur,' the bald eagle (Haliaeetus leucocephalus). Guanlong is the earliest known tyrannosaur (Late Jurassic, 158–163 million years ago) and one of the smallest members of the group at about 4 metres long. Top: Guanlong wucaii © Peter Schouten, reproduced with permission. Bottom: Haliaeetus leucocephalus © Eric Isselée/iStock/Thinkstock.
Chapter opening image: The evolution of flight © Jeff Goertzen/Australian Geographic.

References to websites (URLs) were accurate at the time of writing. Neither the author nor Columbia University Press is responsible for URLs that may have expired or changed since the manuscript was prepared.

For my father
A great friend and inspiration

Contents

Foreword

Towards the end of the twentieth century there was a flurry of publication of dinosaur books, so much so that even professional palaeontologists stopped noticing the new titles. It was therefore with genuine surprise that I read the book in your hands, because I was impressed by how many exciting discoveries have been made in the last decade or so, and how much we have learned about the biology of dinosaurs.

One area in particular – the origin and diversification of birds – has seen an astounding turnover of productive discovery and research. Yes, I was part of many of these discoveries, but it's like watching a child grow up: you don't see the differences from week to week, but one day you are shocked to realise that your youngster is fully fledged!

One tends to think of palaeontology as a field where science marches forward at a slow and ponderous pace. However, the advances of the last decade and a half read like a science fiction novel as palaeontologists have embraced new and exciting technologies and approaches to learn things that I'd never have thought possible when I started collecting dinosaurs professionally many decades ago.

For example, in the early 1990s who would have thought that many astoundingly well-preserved feathered dinosaur and bird skeletons would have been discovered in China and other parts of the world?

Profound thinking had predicted 20 years earlier that some dinosaurs should have been covered by feathers, and artists had even started drawing them with feathers in the 1970s. But the chance of finding preserved feathers in fossils seemed so remote that, even when I saw the first feathered dinosaur in 1996, I was trying to find other explanations for the halo of fuzz around *Sinosauropteryx*.

Now there are thousands of fossils of feathered dinosaurs and birds, from China, Mongolia, and other parts of the world. Even my own backyard, the province of Alberta in Canada, has now produced ornithomimids with feathers, and non-avian dinosaur feathers in amber. These discoveries should be exciting enough, but they have also opened up new areas of research on the origin of feathers, and the origin of powered flight in birds.

Studies of fossilised soft tissues have given us clues about growth, longevity and physiology. Stomach contents in the feathered dinosaurs have revealed diets that include fish, lizards, birds and mammals. The discovery of pigment-holding structures in feathers and skin has even revealed the colours and patterns of the feathered dinosaurs and early birds!

Thoroughly researched, with new interviews, this is one of the best, most accessible dinosaur books that has appeared in years.

– Philip Currie, MSc, PhD, FRSC
Professor and Canada Research Chair, Dinosaur Palaeobiology
University of Alberta

Preface

How I came to write this book.

Most palaeontologists will tell you they arrived in their job because, as a child, they loved prehistoric creatures, particularly dinosaurs. They just never grew up – or at least they never stopped seeing a world filled with wonder and excitement. In that sense, I never grew up either. I never stopped being the little kid staring up in awe at the 32-metre-long fossil of 'Dippy' the *Diplodocus* that fills the vaulted Central Hall of London's Natural History Museum.

I've long been in awe of the museum too, a grand cathedral to the natural world and Darwin's theory of evolution by natural selection, all showcased in architect Alfred Waterhouse's great neo-Romanesque confection of a building. The overall effect is one of Victorian grandeur, but with splendid detailing, right down to the gargoyles and statues of pterodactyls, cats, sabre-toothed lions, wolves, bears and numerous other creatures both prehistoric and modern.

After childhood visits to the museum helped nurture a passion for all things dinosaur, many roads in my early life led back to that imposing Victorian building in London's leafy suburb of South Kensington. The offices where my dad had his business were just around the corner, so as a teen I'd often pay visits, wandering the galleries or sitting out the front on the grass. When I started my degree in biology in 1996, I chose Imperial College, which lies right in the shadow of the museum.

While studying at Imperial, I spent some time volunteering in the museum press office. Then for my undergraduate project in my final year I spent a month in the mammal tower of the museum pulling out drawer after drawer of primate bones and measuring the skulls with calipers (my study didn't come to any strong conclusions about the evolution of primate body size as it was supposed to, but I drank in the opportunity to rummage through the museum collections). The next year I decided to take a master's degree at the museum itself, studying biodiversity, evolution and museum science.

Spending a year at the museum as a postgraduate was when I really grew to love the place. We had pretty much free rein behind the scenes, and I quickly came to realise that only a tiny fraction of the collections is on display. Most of the 70 million or so specimens (some of which were collected by Charles Darwin on the *Beagle* and by Joseph Banks, Captain Cook's botanist on the 1768–71 voyage of the *Endeavour*) are squirrelled away in climate-controlled corridors and towers that sprawl on and on, like a miniature campus. I spent many a productive afternoon exploring the bowels of this fantastic institution.

When my studies concluded, I chose not to pursue the academic route, instead becoming a science and environment journalist. This gave me the opportunity to remain involved with the field I loved, but allowed me to constantly learn new things in many different avenues of science. I never stopped following the latest dinosaur discoveries, and continued to write about them whenever I could – some stories for *National Geographic*, others for *New Scientist*. But it was a feature story summarising all the many streams of evidence that birds are descended from dinosaurs, which appeared in Australia's *Cosmos* magazine, that really allowed me to get stuck into the feathered dinosaurs topic. When this story appeared in an anthology of Australian science writing, and subsequent conversations revealed it was a topic that many people found intriguing but knew very little about, I had the idea to write this book.

Today I live in Sydney, where I'm the editor of *Australian Geographic* magazine. I still love to read, write and talk about dinosaurs whenever I get the opportunity (sometimes I even wear dinosaur pyjamas). Researching and writing this book has not only been fascinating, it's been a lot of fun, and I've relished the opportunity to interview and correspond with many of the world's top dinosaur scientists – in Australia, the United States, the United Kingdom, China and Canada – and hear their incredible stories of discovery first hand.

Returning to the Natural History Museum in mid-2012 to begin the detailed research for this book brought back memories of my studies there more than a decade earlier, as well as my childhood visits. I was there to speak to dinosaur experts, but most importantly to take a look at a fossil replica of the famous 'London specimen' of the 'first bird', *Archaeopteryx*, which is on display in the museum's galleries.

Archaeopteryx, discovered in 1861 in Germany, was the first substantial piece of evidence that birds are the descendants of dinosaurs – and it was where my research for this book fittingly began. When you start to think about the fact that birds really are living dinosaurs, you begin to see everything in a new light.

Introduction:
A whole new world

The revelation that birds are dinosaurs and the discovery of
many stunning fossils of feathered dinosaurs in China has
opened up a window into an unknown prehistoric world.

Imagine, if you will, a world filled with billions of dinosaurs. A world where they can be found in thousands of shapes, sizes, colours and classes in every habitable pocket of the planet. Imagine them from the desert dunes of the Sahara to the frozen rim of the Antarctic Circle – and from the balmy islands of the South Pacific to the high flanks of the Himalayas. The thing is, you don't have to imagine very hard. In fact, wherever you live, you can probably step outside and look up into the trees and skies to find them. For the dinosaurs are the birds and they are all around you. Dinosaurs didn't die out when an asteroid hit the earth 66 million years ago. *Everything you were told as a child was wrong.*

The idea takes some getting used to. On the face of it, birds don't seem that similar to dinosaurs – they're small, bright, quick and covered with feathers, whereas the dinosaurs I was told about as a kid were hefty, lumbering beasts, scaly and reptilian in aspect, much more like a crocodile than a bird. But the clues were there all along if only we knew what we were looking for. Theropod dinosaurs (the bipedal, carnivorous variety) share numerous small features of their skeletons with birds – far more than either share with any other group of animals.

An early clue to the link between theropods and birds came with the discovery of the first fossil of *Archaeopteryx* in a Bavarian quarry in 1861. It has been called the most important fossil ever found, not least because of what it tells us about dinosaurs. Labelled the 'first bird' – *Urvogel* in German – this prehistoric animal had wings and feathers, but also the long bony tail and teeth of a reptile. Its similarity to *Compsognathus*, a small dinosaur found in the same German limestones, was striking, and was even remarked upon at the time by evolutionary biologist Thomas Henry Huxley.

It was only two years since Darwin had unveiled his theory of evolution by natural selection in *On the Origin of Species*, and it seems the world wasn't yet ready for the revelation of the link between dinosaurs and birds. That connection would remain obscured until 1964, when palaeontologist John Ostrom stumbled upon the fossils of several lithe, athletic and deadly-looking dinosaurs called *Deinonychus* in the badlands of Montana. Ostrom resurrected the idea that *Archaeopteryx* was closely related to theropod dinosaurs such as *Deinonychus*, and so began the 'dinosaur renaissance' of the 1970s, which saw leading experts redefine dinosaurs as intelligent, speedy, warm-blooded creatures that were similar to birds.

An earth-shattering find

The idea that birds were the direct descendants of dinosaurs still had its detractors, but much of the opposition fell away in 1996, when the fossil of a little dinosaur from China shook the very foundations of palaeontology. *Sinosauropteryx* was undoubtedly a dinosaur, but the fossil clearly showed that it was covered in a fuzzy down of protofeathers that later studies showed would have been ginger-coloured. This was the first of the feathered dinosaurs to be discovered, but whole flocks of feathered dinosaurs have since burst onto the scene, and we now have evidence for feathers of some kind in about 40 species. Most hail from the Early Cretaceous (100–145-million-year-old) shales of China's north-eastern Liaoning Province, which preserve fossils in remarkable detail.

Every new fossil is a small pebble of proof in an avalanche of evidence confirming that birds are the descendants of the theropod dinosaurs and that these animals were incredibly bird-like. Ever since the 1870s, the hunt for dinosaurs has been a competitive business, filled with adventure and excitement. The days of Edward Drinker Cope and Othniel Charles Marsh, North America's most famous fossil hunters, were the most ruthless of all, culminating in the 'bone wars' that destroyed them both. These days dinosaur fossils have become so valuable – not only to science but also to collectors – that an improbable but thriving trade in fake and illegal dinosaur specimens has become an increasing headache for palaeontologists. This has meant that modern dinosaur experts have to have the penetrating logic of Sherlock Holmes, the adventurous spirit of Indiana Jones and the wisdom of Solomon just to stay ahead of the game.

Beyond confirming the dinosaur–bird link, the fossils have offered clues about how feathers evolved in the first place, and how they might have been used for flashy display purposes and insulation long before they ever helped any creature become airborne. What is most exciting about the latest discoveries, though, are hints at how dinosaurs did eventually take to the skies. We now know that the dinosaurs most closely related to birds were small predatory species, a number of which, such as *Microraptor*, *Anchiornis* and *Xiaotingia* – incredibly – had four wings and a long feathery tail. Their hind limbs and tails had flight feathers of the kind we see only on the forelimbs of modern birds, so it's likely they used them to glide between the trees of China's swampy Cretaceous forests.

We also now know that dinosaurs were bird-like in many other aspects of their physiology and behaviour too. From nesting, brooding and sex, to metabolism, development and even the diseases that afflicted them, many of the traits found in birds today were inherited from the dinosaurs. The boundary between dinosaurs and birds has become utterly blurred.

Bizarre menagerie

The Chinese forests of the Jurassic and Cretaceous would have been filled with a bizarre menagerie of bird-like dinosaurs, which later shared these same forests with birds themselves. Now we even know something about the colour of the feathers of the four-winged species that flitted from tree to tree. While *Microraptor* appears to have had plumage of a deep iridescent blue–black, *Anchiornis* was dappled black and white with a red head crest.

Other feathered dinosaurs were stranger still. Pigeon-sized *Epidexipteryx*, found in the Mid- to Late Jurassic (152–168 million years ago), didn't use its feathers for flight at all; it had a downy covering of fuzz for insulation and four long, ribbon-like feathers that emerged from its tail. This discovery was one of the strongest suggestions that feathers were important as insulation and for display long before they were useful for flight. Weirder still were *Epidexipteryx's* incredibly long fingers – its middle finger was half the length of its entire body. These strange features suggest that the creature scrambled around in the trees and may have used its long digit to skewer fat grubs in treeholes and crevices, just as the aye-aye of Madagascar does today.

Even more surprising perhaps is that such well-known dinosaurs as *Velociraptor* (and its North American counterparts *Deinonychus* and *Dromaeosaurus*) may have had long feathers on their forearms – in effect small wings – to help them work as a pack and bring down larger herbivores. These proto-wings would have provided a bit of lift as the animals ran and leapt into the air, vicious sickle claws outstretched.

Many of the dinosaurs closely related to birds had beaks too, such as *Oviraptor* and *Caudipteryx*, and the loss of teeth in birds is thought to have been one of many adaptations, along with bones full of air pockets, that eventually helped them reduce their body-weight enough to allow for flight. Not all feathered dinosaurs were small animals, though, and fossils revealed in recent years have shown that there were some truly giant bird-like species. These include the parrot-beaked, 8-metre-long *Gigantoraptor*, whose forearms and a tail were probably clad in large feathers, which it used for earth-shaking displays to woo mates, much as ostriches do today.

For a while *Gigantoraptor* was the largest feathered animal known ever to have lived, but that all changed in 2012 with the discovery in China of a bus-sized early relative of *Tyrannosaurus rex* called *Yutyrannus huali* (which means 'beautiful feathered tyrant'). This 1.5-tonne predator was covered in downy filaments of dino-fuzz, perhaps giving it an incongruously fluffy covering similar to that of a chick. This was our first evidence that truly massive dinosaurs sometimes had feathers too, and even hinted that the mighty *T. rex* itself could have been covered in feathers.

Our changing view of dinosaurs

The way our view of what constitutes a dinosaur has evolved in the last two decades can be seen by re-examining Stephen Spielberg's 1993 film *Jurassic Park* (and Michael Crichton's 1990 book from which it was adapted). The film's quick-witted, pack-hunting velociraptors owed much to the picture John Ostrom had painted of *Deinonychus*. I was 14 years old when my dad took me to see the movie on the big screen at the Odeon in London's Leicester Square. I remember the thrill – the fear, the joy, the wonder – of seeing a menagerie of dinosaurs brought back to life on the big screen. Never before had anyone done such a wonderful job of imagining and depicting what these animals might have looked like in life, how they might have moved and behaved, and even what sounds they might have made. In fact, Steven Spielberg's team did such a brilliant job that today, more than two decades after the film was made, the reconstructions still look fantastic, even though we would imagine them somewhat differently today.

But it wasn't just the moving images that inspired and excited me – it was also Crichton's ingenious plot: the idea was that we might be able to find dino DNA in the fossil record and use modern cloning technology to reread the blueprints and bring these long-gone creatures back. He even came up with a seemingly plausible place to find this DNA – inside mosquitoes that had supped on dinosaur blood and become entombed in amber. But would it ever be possible to find enough dinosaur DNA to resurrect them from extinction? In perhaps the most audacious modern dinosaur project, one scientist is looking at reprogramming the development of a chicken to awaken dinosaur traits (such as teeth and a bony tail) lying dormant inside its genes, so that some day we might watch a baby dinosaur hatch from the egg of a chicken.

Along with the new fossils, and renewed interest in dinosaurs, have come fresh interpretations of how dinosaurs lived their lives. In 1993 nobody could ever have predicted that we might know something about the sounds that dinosaurs made and the colours they were decked out in, but clever new methods have begun to probe these kinds of details too. Experts are also starting to pin down some of the reasons why birds survived the comet strike that doomed the rest of the dinosaurs to extinction 66 million years ago.

We have learnt more about dinosaurs in the two decades since *Jurassic Park* than during the whole of history up to that point. The 1990s seemed like a golden age of dinosaur discovery, but fossil finds since then have dwarfed it. Around one new species is currently discovered every week, many in China, but others in South America, Mongolia and Africa. There's so much new knowledge it's hard to keep up, but one thing's for certain – if you love dinosaurs, this a great time to be alive.

If, like me, you can still imagine a prehistoric world filled with wonder and excitement, the best may yet lie ahead of us. If the last 20 years have taught us anything with regard to what we know about dinosaurs, it's to expect the unexpected. Want to know more? Get ready to unthink what you thought you knew and come with me on a journey into the deep, dark depths of the Jurassic…

Before we begin

I want to start by clearing up some common misconceptions about dinosaurs. If you're a dino buff then you'll probably know all of this already and want to skip forward a few pages.

1 What is a dinosaur?

Dinosaurs are a group of reptiles defined by many features of their skeletons – most particularly the fact they hold their limbs erect beneath them rather than out to their sides in a sprawling posture as lizards and crocodiles do. Dinosaurs are made up of two major subgroups: the saurischian or 'lizard-hipped' dinosaurs, which included the giant long-necked sauropods (such as *Diplodocus*) and all the bipedal, predatory theropods (such as *Tyrannosaurus* and *Velociraptor*); and the ornithischian or 'bird-hipped' dinosaurs, which included heavy-set and armoured species (such as *Triceratops* and *Ankylosaurus*) and herd-living herbivores (such as *Hadrosaurus* and *Pachycephalosaurus*). Officially dinosaurs are deemed to be all the animals that descended from the last shared ancestor of the ornithischian and saurischian groups. Confusingly, birds are theropods and are therefore part of the 'lizard-hipped' group.

2 Dinosaurs and people didn't coexist

Modern humans that look like you and me evolved around 200 000 years ago, while our more distant ancestors diverged from chimpanzees 6–7 million years ago. The last of the non-avian dinosaurs (that is dinosaurs that weren't birds) disappeared in a mass extinction around 66 million years ago at the end of the Cretaceous period. At around the time the dinosaurs winked out, the lineage of mammals that eventually gave rise to humans and all other primates included a small squirrel-like animal called *Purgatorius*.

3 Geological history is *vast*

The earth formed around 4.5 billion (4500 million) years ago. A common analogy is to imagine the earth's history as a 24-hour clock. On this scale, the earth formed at midnight. Simple bacterial life appeared at 6 am, more complex plants and animals didn't make it onto land until 10 pm and humans appear at one minute to midnight. Dinosaurs enter the stage at around 11.20 pm, in the Triassic period (200–250 million years ago), enjoy great success through the Jurassic (145–200 million years ago) and make their exit at 11.50 pm at the end of the Cretaceous period (66–145 million years ago).

4 Birds are dinosaurs

Birds evolved from within a group of fast two-legged predatory dinosaurs called theropods, which includes such creatures as *Velociraptor* and *T. rex*. This means that dinosaurs are in fact more successful today than they have been at any other point in their history. There are nearly 10 000 living species and perhaps as many as 400 billion individual birds flitting about on the planet at any one time. As British writer Colin Tudge says, 'The dinosaurs did not disappear when the putative asteroid struck, 65 million years ago. They are calling to us from every hedgerow'. One estimate of total dinosaur diversity (including those we haven't yet found as fossils) puts it at around 1850 species – so, by sheer weight of numbers, the small, feathered, flying dinosaur model really has been fabulously successful. Flight opened up a whole new range of lifestyles for dinosaurs and it was a leap into the skies from which they've rarely looked back. Living mammal species number only about 5500, and birds outnumber them and all other groups of terrestrial vertebrates, including amphibians and reptiles.

5 Ancient marine reptiles and flying pterosaurs are *not* dinosaurs

The flying pterosaurs and the aquatic ichthyosaurs, plesiosaurs, pliosaurs and mosasaurs were large prehistoric reptiles that lived at the same time as the dinosaurs, but they were not dinosaurs. Pterosaurs were the sister group to dinosaurs and birds, and in turn all of these animals are close relatives of crocodiles (which are therefore the closest living relatives of birds). The other ancient marine reptiles are more distantly related members of the reptile group, which also includes turtles, snakes and lizards.

6 What do I mean when I say 'dinosaur'?

Since birds are simply small, specialised, mostly flight-capable forms of dinosaur, this presents a problem for how to define the rest of the dinosaur group. Officially, everything we would traditionally have thought of as a dinosaur is a 'non-avian dinosaur', but that's one ugly mouthful, so in this book you can assume that when I say 'dinosaur' that's usually what I mean.

7 What defines a bird?

It used to be easy to define a bird: birds had feathers and beaks, they flew, they were bipedal, they laid eggs, they were warm-blooded and quick-witted, and they had wishbones. The problem is that most of the 30 or so once-defining features have quite literally gone the way of the dinosaurs. Of course, birds still have all these things, but now we know that many dinosaurs had them too. So what now defines a bird? In some ways the distinction between birds and dinosaurs is arbitrary. Mike Benton, a palaeontologist at the University of Bristol in the United Kingdom, says that powered flight and a wing that was large enough to achieve it, is the one remaining thing that separates *Archaeopteryx*, the 'first bird', from its dinosaur relatives, but it may be only a matter of time until other dinosaurs are found that had powered flight too.

8 Dinosaurs didn't all live at the same time

Compared to the very brief tenure of humans on this planet, dinosaurs ruled the roost for an *extremely* long time. The earliest recognised dinosaur is a bipedal labrador-sized animal called *Nyasasaurus*, which was named after Lake Nyasa in Tanzania (today called Lake Malawi). The fossil was collected in the 1930s but languished for many decades in a drawer at London's Natural History Museum. A reappraisal of it in 2012 found it to be up to 247 million years old, making it significantly older than any other known dinosaur (assuming it really is a dinosaur – the remains are very incomplete). From these somewhat diminutive beginnings in the early Triassic, dinosaurs went on to great peaks and troughs of diversity in the late Triassic and through the Jurassic, before slowly declining in species numbers towards the end of the Cretaceous, when something cataclysmic finished them off 66 million years ago. A 180-million-year run really isn't a bad innings. During this period, individual dinosaur species were continuously going extinct while other new ones were evolving.

9 Birds and dinosaurs *did* live at the same time

The first bird, or at least one of its fairly close relatives, was an animal called *Archaeopteryx*, which lived in the Late Jurassic around 150 million years ago. Non-avian dinosaurs didn't disappear until 66 million years ago, so we know that birds and dinosaurs coexisted for at least 85 million years. For a long time we had very few early bird fossils, but large numbers of exquisitely preserved fossils from China are now showing that a diverse assemblage of dinosaurs and birds shared the same habitat there during the Early Cretaceous period.

Flying Dinosaurs

1

The missing link

How the connection between dinosaurs and birds was found – and then lost again for 50 years.

Our story begins in the Altmühl Valley in Bavaria, Germany. It looks nothing like Bavaria now, for we have stepped back 147 million years into the past, to the Late Jurassic period, and the land here is part of a subtropical island archipelago much nearer to the equator than today. An ominous-looking afternoon thunderstorm has rolled in and gravid clouds are pelting the shallow lagoons with a torrent of tepid raindrops.

The warm waters teem with jellyfish, crustaceans, corals and other invertebrates, as well as fish, crocodiles and large marine reptiles such as ichthyosaurs. Swooping through the skies above are pterosaurs, flying reptiles related to dinosaurs. Along the shoreline a pack of *Compsognathus* – turkey-sized carnivorous dinosaurs – fluff up their downy feathers and shake themselves in the rain before chirping and darting into the thick cover of the woodland fringe.

Another reptile here can fly – or at least a reptile of sorts, for it looks very much like a modern bird. About the size of a raven or a crow, this half-metre-long creature has wings and a feathered tail that seems unusually long for its body. Its flight feathers are pale with dark tips, giving it a pattern something like a magpie. Unlike any modern bird, however, it has a long bony tail, claws on the end of its wings, and teeth. This animal is *Archaeopteryx* and its species is one of the first birds truly capable of powered flight.

Though *Archaeopteryx* can take to the skies, it has little of the aerial aptitude of a modern bird. Taking off costs it significant energy and effort, and it is prone to ungainly crash landings. The extra weight it carries from its bony tail is only part of the problem, as it also lacks a keeled sternum (or breastbone) and the wing-powering mass of muscle that attaches to this bone in modern birds.

In other respects, though, *Archaeopteryx* is already attuned to flight. Relative to body size, its brain is large, and its cerebral hemispheres and regions used for vision and movement are well developed, giving it a keen sense of sight and spatial awareness. On its hind limbs are feathers that help stabilise it and act as air brakes. This allows the otherwise cumbersome creature to turn more adeptly and fly steadily, so it can negotiate the dense wooded environment; swoop out of the reach of predators; and sneak up on the lizards, frogs, beetles, dragonflies and other insects it feeds upon in these prehistoric forests.

The lagoons here sometimes turn stagnant in the heat, and the thick, silty mud that carpets their floors is perfect for preserving – in great detail – the remains of any dead animals that sink into them. Over the millennia many *Archaeopteryx* do meet their end on the floor of these lagoons, and in the resulting rocks they would surely have remained if not for a phenomenal turn of fate that led Bavarian quarrymen to chance upon some of their delicate imprints 147 million years later.

The first bird

The first specimen of *Archaeopteryx* was discovered in Germany, in a slab of Bavarian limestone in 1861, just two years after the publication of Charles Darwin's *On the Origin of Species*. It is one of the most important fossils ever found. Puzzlingly, the animal preserved in the rock shared many features with theropod dinosaurs but was unmistakably a bird, with perching feet and well-developed flight feathers.

Archaeopteryx is a transitional fossil with a mixture of features found in both birds and reptiles, says Paul Barrett of London's Natural History Museum, one of the United Kingdom's foremost dinosaur experts. 'No modern bird has teeth, and no modern bird has a bony tail; both of these are reptilian features. Although it is known definitively as the first bird, early on it would have been recognised that it has these reptilian features, and this was a kind of smoking gun.'

Only 11 specimens (and a feather) have ever been found, and the Natural History Museum still has the one first used to describe the species by its then director Professor Richard Owen in 1863. Scientists come from across the world to study the 'London specimen', and it still plays a central role in today's debates about the relationship between early birds and feathered dinosaurs.

Many of the other specimens are in German institutions, including the 'Berlin specimen', which is the most beautiful of all. One is in the United States in a private institution. Another is in private hands, presumed lost. One was misidentified as a pterosaur and re-identified as *Archaeopteryx* by palaeontologist John Ostrom in the 1970s. 'There aren't that many of them, but between all of them we have a fairly good idea of what the skeleton looked like', Barrett says. 'They are different-sized individuals and slightly different ages. It's been suggested that one or two of them might represent a different genus of bird that's closely related to *Archaeopteryx*, but that's contentious.'

3

Before the discovery of *Archaeopteryx*, scientists had already realised there were skeletal similarities between birds and reptiles such as crocodiles. But when Darwin published *On the Origin of Species* in 1859, no transitional fossils had been found. Today, after several centuries of collecting, the fossil record is surprisingly patchy, but in Darwin's day it was even more problematic to use it as evidence for the kind of gradual changes through evolution that he proposed. Instead he looked to living creatures, such as the multitude of closely related Galapagos finches, to make his points more forcefully.

In one chapter of his work, where he methodically laid out the flaws in his theory, Darwin noted:

> But, as by this theory innumerable transitional forms must
> have existed, why do we not find them embedded in countless
> numbers in the crust of the earth? ... I believe the answer
> mainly lies in the record being incomparably less perfect
> than is generally supposed ... The crust of the Earth is a vast
> museum; but the natural collections have been made only at
> intervals of time immensely remote.

Then in 1860, with impeccable timing, the fossil of a Jurassic-era feather turned up in a limestone quarry in the Solnhofen region of Bavaria's Altmühl Valley. These fine limestones had been used for building materials and paving slabs as early as Roman times, but had come into their own in the 1700s with the development of lithography, a printing technique that used slabs of the stone engraved with acid. Lithography required only the finest, smoothest slabs, and relied on quarrymen carefully splitting the rock by hand in search of them – the fossils they found of fish, winged insects, shrimps and even pterosaurs were simply a lucrative sideline. The species the feather belonged to was named *Archaeopteryx lithographica* (literally 'ancient wing, written in stone') by German palaeontologist Hermann von Meyer.

The following year a largely complete skeleton, minus the skull, of what appeared to be a puzzling crow-sized bird was given to doctor Karl Häberlein by a sickly quarryman in exchange for medical attention. Squabbling between German bidders led Häberlein to sell the specimen, along with the rest of his extensive fossil collection, to Richard Owen, who paid £700 (about AU$120 000 today). Häberlein, an elderly widower, is rumoured to have used the money to cover a dowry and wedding for his daughter. Though spending such a sum of money seemed unconscionable to the trustees of the museum, it now seems a small figure to have paid to secure what is the institution's most valuable piece.

Archaeopteryx and the evolution debate

Owen was one of the foremost palaeontologists and anatomists of his day, and was a well-known public figure; he devised the term 'dinosaur' itself in 1842, meaning 'fearfully great, a lizard'. He was also a darling of London's Victorian elite and gave the children of Queen Victoria herself biology lessons. Nevertheless, he was a 'lean-faced, pop-eyed man of pedantic turn of phrase ... notorious for using ridicule and malicious attacks to further his own position', says anthropologist and writer Pat Shipman.

A man with deep religious convictions, Owen didn't like Darwin's idea of natural selection one little bit. As biologist Thor Hanson notes, Owen had built his career:

> on the firm belief that species were created and altered only
> by the hand of God. If *Archaeopteryx* could be perceived as
> an intermediate step between reptiles and birds, it would be
> dangerous fodder for Darwinists. They would call it evidence
> that birds and their most distinctive feature, feathers, evolved
> from reptiles.

Within months of receiving the *Archaeopteryx* fossil in London in 1862, Owen had dashed off a paper in which he made no mention of a missing link, describing the fossil as 'unequivocally a bird' and passing off its reptilian features, such as a tail and teeth, as 'a closer adhesion to the general vertebrate type'. The scene was now set for a battle of wills with his great intellectual rival, the staunch Darwinist Thomas Henry Huxley.

History remembers TH Huxley as 'Darwin's bulldog', and he was certainly a vocal champion of the idea of evolution by natural selection, as well as a close friend of Darwin. Huxley was a great evolutionary biologist and comparative anatomist. He was based at the Royal School of Mines, now one of the constituent schools of Imperial College. In contrast to Owen's noted pomposity, Huxley was apparently witty, well liked and a gifted speaker.

In 1868 he published his own paper on *Archaeopteryx*, detailing the remarkable similarities of the fossil to the small dinosaur *Compsognathus* and arguing that the species was not only the first bird but also a very clear missing link between birds and dinosaurs. He later noted:

> Birds are modifications of the same type as that on which reptiles are formed, and if this similarity of structure is the result of community of descent, we should expect to find, in the older formations, birds more like reptiles than any existing bird, and reptiles more like birds than any existing reptile. If the geological record were sufficiently extensive ... we ought to find an exact series of links, but this, of course, is hardly to be expected.

(Huxley surely couldn't have imagined that with all the recent feathered dinosaur discoveries in China, this 'exact series of links' is precisely what we have found.) In his paper on *Archaeopteryx*, he also went on to include a withering rebuke of Owen for confusing the left and right leg of the specimen and getting the pelvis the wrong way round.

Huxley recognised that the hind parts of bird and dinosaur skeletons are strikingly similar. 'If the whole hind quarters, from the ilium [pelvis] to the toes, of a half-hatched chicken could be suddenly enlarged, ossified, and fossilised as they are', he wrote, 'they would furnish us with the last step of the transition between birds and reptiles; for there would be nothing in their characters to prevent us from referring them to the Dinosauria'.

Similarities between *Archaeopteryx* and *Compsognathus* – fossils of which are found in the in the same Solnhofen deposits – included the fact that both animals were bipedal, had an upright stance and shared numerous features of their ankle joints. Bipedal – two-legged, rather than four-legged – animals are few and far between, and making the transition requires a complex suite of adaptations, so it has happened rarely in evolutionary history. Full bipedalism has only evolved a handful of times in lineages of modern animals – in kangaroos, rodents, birds and the line leading to humans. Making the switch to two legs also frees up the forelimbs for other purposes – such as making tools or flapping to get into the air – and has therefore been associated with great evolutionary success, but that's something we'll come back to later.

Referring to the similarity between the bipedal nature of birds and the dinosaurs *Iguanodon* and *Megalosaurus*, Huxley noted:

> When we consider what a very strong piece of evidence this is, we are forced to the conclusion that the evolution of birds from reptiles, by some such process as these facts indicate, is by no means such a wild speculation as it might, from a priori considerations, have been supposed to be.

'Huxley noticed a lot of similarities between skeletons of things like pigeons and small meat-eating dinosaurs', Paul Barrett says.

'He proposed that birds and dinosaurs were very closely related, on the basis of things like the similar hollow spaces within the bones and similar bone structures in general.'

The degree of similarity was such that one of the later specimens of *Archaeopteryx*, found in 1951, was initially mistaken for a young *Compsognathus*. Yale University's OC Marsh (see chapter 3) would endorse Huxley's ideas about bird origins when he described fossils of the early toothed birds *Hesperornis* and *Ichthyornis* in 1880. Despite this, the idea of birds evolving from dinosaurs didn't gain traction at this time, Barrett says. 'This whole debate just doesn't really take off, rather oddly. People are still busy describing fossils and just naming things and getting them out of the ground. The idea was never really taken up and then fell out of favour. Bit of a shame, because Huxley was dead on with most of the stuff.'

The Berlin specimen and dinosaur death

The next specimen of *Archaeopteryx*, found in 1877 in a quarry near Eichstätt – also in the Altmühl Valley – was a complete skeleton with a skull and its wings splayed out to the sides. The fossil, which ended up at the Naturkunde Museum in Berlin and came to be known as the 'Berlin specimen', is one of the most beautiful fossils ever collected. Nearly every delicate bone is visible, and it has a great fan of feathers marked in the limestone around the body and its long bony tail.

'The Berlin specimen is truly remarkable – and beautiful', John Ostrom wrote in a 1975 article for *Discovery* magazine:

> The skeleton is almost 100 per cent complete, with clear feather imprints and (unlike the London specimen) a very bird-like skull with teeth! The head and neck are arched back over the back: the long reptile-like tail is stretched out behind; the long bird-like legs are doubled back slightly; and

the very long dinosaur-like arms and hands, with their great span of feathers, reach out on either side.

The London specimen had been headless, but the Berlin specimen was replete with every detail. Owen had predicted that *Archaeopteryx* had a beak, like a bird, but just as Huxley had suggested, it also had a full set of tiny pointy teeth, much like its close relative *Compsognathus*.

I've spent some time looking at the shape of the fossil and trying to imagine what this strange crow-sized bird might have looked like in life. As I pondered its shape, something began to perplex me: why was it twisted into such a strange position? If you've ever seen one of the *Archaeopteryx* fossils in a museum or even just a photo (see image section), you may have noticed the unusual shape of the body. All the known specimens display a typical 'death pose' also seen in many other dinosaurs fossils – the head and neck are awkwardly twisted back over the body and often the tail is recurved back to meet the head.

'Virtually all articulated specimens of *Archaeopteryx* are in this posture, exhibiting a classic pose of head thrown back, jaws open, back and tail reflexed backward, and limbs contracted', says Kevin Padian of the University of California, Berkeley. This position is known as opisthotonus (Greek for 'tightening behind'). The cause has been the topic of some debate since the 1920s: some experts thought it might be caused by water currents moving the neck after death, or that it was simply the result of rigor mortis or drying tendons.

But in 2007, Padian and veterinarian-turned-palaeontologist Cynthia Faux, of the Museum of the Rockies in Bozeman, Montana, came up with an altogether different proposal. Faux told reporters that the traditional explanations made no sense to her: 'Palaeontologists aren't around sick and dying animals the way a veterinarian is … [we] see this posture all the time in disease processes, in strychnine cases, in animals hit by a car or in some sort of extremis.'

Instead, her gruesome explanation is that the position is caused by the agonised death throes of animals undergoing severe trauma to the central nervous system, as might be caused by asphyxiation, brain damage and a slow and painful death. Opisthotonus in living (or rather dying) animals can be caused by suffocation, meningitis, tetanus or poisoning – in short, anything that damages the cerebellum (hindbrain), causing muscles that normally help with balance to go in to overdrive, clenching at full force. Faux tested the idea with a series of birds of prey that had not undergone this kind of trauma before their death. She watched their decomposition over a period of time, and found that none of their heads curved back over their bodies.

The finding was an interesting one because it meant that fossils preserved in this position were also saying something about the way the animals had died. Furthermore, it seemed to corroborate with the idea that some fossils – particularly specimens of *Archaeopteryx* from Solnhofen and feathered dinosaurs from Liaoning in China, such as *Sinosauropteryx* and *Caudipteryx* – had been poisoned by volcanic gases or suffocated during a rain of ash. It seemed the case was closed, but few things are ever this simple in the world of science: hypotheses are made and then broken, tested and retested, sculpted and adapted.

Shortly thereafter, two groups of scientists – in Switzerland and Ohio – simultaneously and coincidentally came up with fresh evidence that the pose was the result of something that happened after death, during decomposition. In February 2012 Achim Reisdorf, a doctoral student at the University of Basel, pointed out that for an animal to be fossilised in a position that occurred before death, it would have to be preserved in the same spot almost immediately, which is considerably unlikely given many fossils of land animals were formed after their body was carried into water and settled on the floor of a river, sea or lake before being buried with sediments.

Instead, Reisdorf was convinced the death pose might have

something to do with the water itself. Along with German palae-ontologist Michael Wuttke, he bought a series of chicken necks and watched what happened to them after they were left in buckets of water. Rapidly and spontaneously, the necks bent back by 90 degrees. When left for much longer – up to three months – some bent back up to 140 degrees. To find out why, the scientists removed skin and muscle from the chicken necks. They discovered that ligaments between the vertebrae were responsible; if these were cut, the necks fell back to their original position. Scientists at Brigham Young University in Utah announced very similar findings just a few months earlier at a meeting of the Society of Vertebrate Paleontology.

In a paper on the find, Reisdorf and Wuttke suggested that strong neck ligaments that were 'preloaded' were essential for all long-necked dinosaurs with a long tail, to provide the appropriate balance. However, following death and immersion in water, the stored energy was powerful enough to arch the spine back, a phenomenon that increased with decay. This seems to make a lot of sense. *Archaeopteryx* and many of the other feathered dinosaurs are animals with long bony tails – *Sinosauropteryx* has the largest number of tail vertebra of any dinosaur. Perhaps the death pose is a consequence of the neck ligaments necessary to counterbalance a long tail.

The wishbone problem

Huxley's theory that birds were descended from dinosaurs really died a death – at least for a time – when Danish amateur zoologist and natural history artist Gerhard Heilmann published an influential book called *The Origin of Birds* in 1926. In this he argued that because dinosaurs don't have collarbones, birds couldn't be descended from them, as their wishbones are made of fused collarbones. Collarbones were known in earlier reptiles that were assumed to be the ancestors of dinosaurs, and Heilmann subscribed to a view of evolution (with the catchy name 'Dollo's law of irreversibility') which held that once a complex trait was lost it was gone for good. Instead, Heilmann suggested that birds evolved from another group of tree-climbing reptiles that were related to dinosaurs but distinct from them, and his ideas held sway for nearly 50 years.

There was actually some merit in the idea that because dinosaurs didn't have collarbones, birds couldn't be descended from them, but what Heilmann would never know was that dinosaurs had already been found not only with collarbones, but collarbones fused into full wishbones.

'The irony is that in 1923 the first *Oviraptor* specimen was found and that specimen very clearly has a wishbone', says dinosaur scientist Phil Currie, a veteran fossil hunter at the University of Alberta, Canada. *Oviraptor* was discovered in Mongolia by a team led by Roy Chapman Andrews, but its wishbone was misidentified when Henry Fairfield Osborn of New York's American Museum of Natural History (AMNH) described the species in 1924.

'Even though more wishbones were found over the years, everybody thought dinosaurs didn't have wishbones, so they were often identified as something else', Currie says. These small bones just don't fossilise well, and were sometimes thought to be lower ribs or were otherwise misinterpreted when skeletons were reconstructed. Today we know that many theropods – including allosaurs, dromaeosaurs and even *T. rex* – had collarbones fused into full wishbones exactly as birds do today. Wishbones, therefore, are not a bird trait, but a piece of dinosaur history that first evolved more than 150 million years ago.

For a century following the discovery of *Archaeopteryx*, few other fossils directly hinting at the transition between dinosaurs and birds emerged. In the early 1970s Polish expeditions to Mongolia co-led by palaeontologist Halszka Osmólska found dinosaurs that had not only a wishbone but also a bony breastbone, as birds do. Soon after, Russian palaeontologist Sergei Kurzanov found a fossil of a bipedal beaked dinosaur in Mongolia's Gobi Desert that bore so many similarities to birds he named it *Avimimus portentosus* ('the amazing bird mimic'). At the time he argued that the 70-centimetre-tall species had quill knobs – pits in the bones where the feathers of living birds are attached with ligaments – but there was no other evidence at that time for feathered dinosaurs, so the discovery was largely overlooked.

Kurzanov was proved right in 2007, when a team of scientists including Mark Norell at the AMNH showed that a specimen of *Velociraptor* excavated in Mongolia in 1998 had clear quill knobs pitted into its forearm. 'A lack of quill knobs does not necessarily mean that a dinosaur did not have feathers', team member Alan Turner told reporters. 'Finding quill knobs on *Velociraptor*, though, means that it definitely had feathers. This is something we'd long suspected, but no one had been able to prove.'

'The more we learn about these animals, the more we find that there is basically no difference between birds and their closely related dinosaur ancestors like *Velociraptor*', said Norell. 'Both have wishbones, brooded their nests, possess hollow bones, and were covered in feathers. If animals like *Velociraptor* were alive today our first impression would be that they were just very unusual looking birds.'

Gerhard Heilmann's ideas about birds descending from an earlier lineage than dinosaurs became the popular orthodoxy until 1964, when an insightful young Yale University palaeontologist made a discovery in the Cretaceous sandstones of Montana that would change everything.

The dinosaur renaissance

Late one afternoon in August 1964, John Ostrom was prospecting for fossils on a slope in central Montana with his assistant Grant E Meyer when, as the *New York Times* puts it, they 'came upon a macabre sight: large and sharp claws reaching out of an eroded mound ... They uncovered the rest of a powerful, three-fingered grasping hand and then a foot. The inner toe stuck out like a sharply curved sickle'. Ostrom later recounted that they had 'both nearly rolled down the slope' in their rush to reach the spot.

Analysis of the fossils back at Yale's Peabody Museum in New Haven, Connecticut, showed them to be from several 125-million-year-old carnivorous dinosaurs, which Ostrom believed attacked their prey by leaping and slashing it open with their lethal sickle-shaped claws. Such fleet-footed and active animals almost certainly had to have high metabolic rates and be warm-blooded, which seemed in stark contrast to the view of dinosaurs as lumbering, dimwitted creatures, which had largely prevailed since Richard Owen first described the group in 1842.

Over the next two years, Ostrom and his team returned to the site several times, collecting more than 1000 bones from at least three individuals. In 1969 Ostrom published his findings, naming the species *Deinonychus* ('terrible claw') and laying out his ideas that it was an active warm-blooded predator, not a plodding cold-blooded reptile. 'This interpretation stimulated a polarizing debate among scientists over the revolutionary idea that at least some dinosaurs, like *Deinonychus* and related *Tyrannosaurus rex* and *Velociraptor*, had more in common with mammals and birds than with ordinary coldblooded reptiles', wrote the *New York Times*. 'This in turn led to the revival of a 19th-century hypothesis that dinosaurs were direct ancestors of today's birds.'

'John [Ostrom] found this beautiful little dinosaur – a fantastic little meat eater that is very, very bird-like in its anatomy. This got him thinking about warm-blooded dinosaurs again', says Phil Currie. '*Deinonychus* and John's discovery of it was the reason the dinosaur renaissance started in the 1970s. In 1973 John published a paper which was more or less the same paper on the origin of birds that had been published more than 100 years earlier by Thomas Huxley.'

Ostrom's ideas about *Deinonychus* are really what inspired the infamous velociraptors in *Jurassic Park*. *Velociraptor* is a closely related but slightly smaller species from Mongolia – the animals depicted by Spielberg were more like *Deinonychus* in size. At the time Michael Crichton wrote his book, *Deinonychus antirrhopus* was sometimes classified in the genus *Velociraptor* as *Velociraptor antirrhopus*, hence the confusion.

The voracious pack-hunting behaviour shown by *Velociraptor* in *Jurassic Park* was actually based on the discovery of several 2-metre-long *Deinonychus* specimens found in Montana around the skeleton of an 8-metre prey animal, a *Tenontosaurus*', wrote Australian palaeontologist John Long in his book *Feathered Dinosaurs*. Generally this association of several predators and prey is seen as 'direct evidence that the raptors were pack hunters that worked together, like modern wolves, to bring down the bigger plant-eating dinosaur'.

Huxley had pointed to anklebones as the key similarity between dinosaurs and birds, but Ostrom also saw a shared history in the wrist joints of birds and these wickedly powerful little carnivores. In an effort to test the link between the two groups of animals, he went on a mission to examine every known *Archaeopteryx* specimen, his careful study of which pointed to a total of 22 skeletal features shared by dinosaurs and birds and no other groups of animals (today more than 80 shared skeletal features have been found). He argued that, except for the feathers, *Archaeopteryx* would look very much like a small dromaeosaur, similar to *Deinonychus* or *Velociraptor*.

Ostrom made the second major discovery of his career in 1970, when he was examining a partial fossil of a pterosaur at a museum in Haarlem in the Netherlands. It didn't look right to him, and when he found out it was from the same Bavarian quarries that had yielded the fossils of *Archaeopteryx*, he realised this fossil was much more significant than had been supposed: 'to my surprise, [I] recognised that a fragmentary specimen displayed in a Dutch museum was not a pterosaur, as it was labelled, but actually was a fifth specimen of *Archaeopteryx*'.

Sadly, Ostrom died in 2005 following a battle with Alzheimer's disease, but Phil Currie recounts one occasion when Ostrom came to visit him in Canada and he showed him some fossils from China. One of the dinosaurs they'd found was a little *Troodon*, which are, relatively, the dinosaurs with the largest brains. 'We found a beautiful specimen in China in the late 1980s and this specimen had been prepared in Canada and I had it in my office', Currie says. 'I told him I had something to show him and I opened up my specimen cabinet, pulled out the drawer and he looked down. Then I got a big shock because I thought I'd given him a heart attack. He sat down so heavy and he was just astonished and he had this look on his face. It was unbelievable. He couldn't even catch his breath.'

Ostrom was shocked because the specimen was remarkably complete and lifelike. It was 'curled up like it had gone to sleep, and with all the bones and this very birdlike pose and ... a long tail that was curled around its body'.

Ostrom's suggestion that these dinosaurs had been warm-blooded was highly controversial in the 1970s, and is still the subject of some debate today, although most palaeontologists now accept that at least the lineage of theropod dinosaurs leading to birds was warm-blooded. In the 1970s, critics of Ostrom's ideas questioned why fossil skin impressions suggested dinosaurs had scaly skin and not fur or feathers for insulation if they were warm-blooded. This led Ostrom on a several-decade-long struggle to prove dinosaurs were feathered – a journey that was to end in spectacular fashion in 1996, when one of the most important fossils of the 20th century was pulled out of the ground in China's Liaoning Province, several hundred kilometres north-east of Beijing.

2

A feathered revolution begins

How a fuzzy little dinosaur from China
turned palaeontology on its head.

In August 1996, when a young farmer in his 30s, Li Yinfang, pulled the fossil of an unusual chicken-sized dinosaur from the ground near the village of Sihetun in north-eastern China's Liaoning Province, he had no idea the specimen would be one of the greatest fossil finds of the 20th century – one that would ignite a furious debate and ultimately confirm that birds are descended from dinosaurs. Liaoning had already yielded many Early Cretaceous fossils of remarkable quality, particularly early birds preserved in exquisite detail – and this specimen was no exception.

One of these birds, *Confuciusornis*, had been discovered nearby just a few years earlier, and quarrying operations set up by poor local farmers had turned up hundreds of specimens. In the years following the discovery of the little dinosaur, the myths around it have evolved. This is in part because prospecting for fossils and selling the specimens on to dealers is illegal. Some versions of the tale recount that Li was digging holes for trees to reforest a hillside. 'It's all very romantic but just not true', says Phil Currie, adding that the site was a large quarry cut into a hillside by the time he visited the following year, and it had been created specifically for the purpose of finding fossils.

Experienced in preparing fossils encased in siltstone, Li split the 60-centimetre piece of rock down the middle to create a slab and counter slab. Each side had a mirror-image impression of a dainty creature. The animal had a relatively large skull, similar in shape to a bird's, a very long tail, and what appeared to be fuzzy fringe of dark feather-like structures along its head, back and tail. Next to its left foot a small fish was also embedded in the rock.

Li, now a guide at Sihetun's dinosaur museum, realised the fossil wasn't *Confuciusornis* and later told Currie it had reminded him of pictures of *Archaeopteryx* he'd seen in schoolbooks. 'I was very surprised because it was very different from the small fish and insect fossils we normally find here. But as ordinary people, we didn't understand it', Li told the *China Daily* newspaper in 2009.

Having no idea of the magnitude of what he had discovered, Li sold one half of the fossil to the Institute of Geology and Palaeontology (IGP) in Nanjing and the other to the National Geological Museum in Beijing, the latter for around US$700 – a vast amount to a poor Chinese farmer, but what seems an absurdly small sum now, considering the fossil's true significance. On the basis of the feathery fringe, Ji Qiang, then director of the museum, initially thought the find was an early bird, but colleagues argued it was a dinosaur.

Currie was the first Western scientist to see the specimen and was instrumental in bringing it to wider public attention. By coincidence, he'd wrapped up a trip to Mongolia's Gobi Desert in September 1996 and then headed to Beijing to spend a few days catching up with colleagues. While there, he was shown a local newspaper clipping detailing a tiny new carnivorous dinosaur that purportedly had feathers. He discounted the idea at first, thinking the markings were more likely to be mineral stains on the rock or fungal growth, but decided to meet with Ji Qiang and take a closer look at it anyway. 'Looking at the newspaper article – it was a crappy photograph on newsprint, and it's not so impressive – and I'm thinking well, what's the likelihood of this ever happening in my lifetime? I think virtually none', he says. 'So I kind of dismissed the idea that they were feathers. But I really wanted to see the specimen anyway, because it was a complete small theropod, and that was very evident from the newspaper photograph.'

Setting up a meeting with Ji Qiang took some organising, and several days went by. When Currie eventually arrived at the museum, he knew something was up because instead of going to the collections area with the scientists, he was taken to a room packed with Chinese media – 'cameras, reporters, photographers – the whole works'. They brought in handmade silk box after box of beautiful specimens from Liaoning: insects and lizards and finally a mammal covered with fur.

'But I wasn't seeing what I came for and I was beginning to think that, well, maybe they decided not to show it to me after all', he says. 'And so just when I was lulled into this feeling that I wasn't going to see it, then they bring out a box, they open it up and there it is. I mean, it was unannounced. It was just there in front of me; they opened it, and within milliseconds I believed in feathered dinosaurs.'

Currie's initial few moments with the first known feathered dinosaur were a magical experience: 'It was almost a jolt because I'd kind of lulled myself into thinking that they weren't really feathers, and therefore I'd thought, "Don't get excited, forget about it. Just get excited about the specimen". And the specimen was really cool, but there's no question that when I saw those feather-like structures around the outside, then it suddenly came back to me that that's really what all the fuss is about ... The detail of the fossil was fantastic and the fact that it was so complete was absolutely stunning. It was one of the finest specimens I'd ever seen in my life.'

Currie quickly realised there was no way the filaments around the fossil were mineral crystals or fungus. They were filaments on the outside of the body and they were independent: 'They were doing all the things that feathers from the [fossilised] feathered birds of that area do, and there was that kind of detail in all the right places.'

A tremendous number of feathered bird fossils that were contemporaries with the dinosaurs had been found in Liaoning, so it seemed like a likely place to turn up a feathered dinosaur. Despite this, nobody had really believed it would ever actually happen.

The find of the century

Sinosauropteryx prima ('first Chinese lizard wing') had already been described in a Chinese publication that had escaped Western interest until then. *Sinosauropteryx*, a bipedal predator related to *Compsognathus*, had a chicken-sized body and was just 1.3 metres from head to tail. The tail had 64 vertebrae, the most known for any predatory dinosaur, making it exceedingly long.

Some in China attempted to bar publication of images of the fossil, perhaps fearing Western scientists would take the credit for it, but word was already out. Currie and his palaeontological colleague from the IGP, Chen Peiji, arrived a few weeks later at the annual meeting of the Society of Vertebrate Paleontology at the AMNH in New York City armed with photos of the dinosaur. These they showed to excited colleagues around the sidelines and in the corridors of the meeting.

There was already a buzz when the meeting started, as the impromptu press conference in Beijing in which Currie had taken part had been widely publicised. 'After I got back, I realised that the news had spread from China to Japan because I started getting phone calls from Japan', he says. 'And then a couple of days before the meeting, I started getting phone calls from Britain because they'd heard about it too ... And then, of course, right at the time that Chen Peiji and I were in New York, him with the photographs, then the whole thing hit the

North American news and the front page of the *New York Times.*'

The next day, on 19 October 1996, the *Times* reported: 'News of the little dinosaur's discovery spread quickly yesterday at the annual meeting ... Although there were no formal announcements of the discovery, scientists crowded hallways and meeting rooms to look at photographs of the Chinese specimen.' Normally at scientific conferences, important discoveries are announced in presentations and lectures, sometimes with media fanfare. 'Rarely are scientific findings of this possible importance presented so casually', the newspaper added. The story featured an illustration of a fluffy-looking *Sinosauropteryx* by palaeontological artist Michael Skrepnick, which helped generate even more interest in the discovery.

'The lines had been drawn in the sand a long, long time ago in terms of what people believed, and a lot of people were really sitting on the fence and one way or another just waiting to see if there was some kind of evidence pro or con to support the idea that birds came from dinosaurs', Currie says. The fossil was therefore a very welcome discovery to those who had first developed and supported the dinosaur–bird link decades earlier.

John Ostrom simply said he was in a state of shock – scientists infrequently get the chance to be so entirely vindicated. 'It was hard to tell sometimes how [John] was reacting, but in this case I could tell he was pretty excited', Currie says.

This little fossil caused a sensation around the world and shook the palaeontological community to its very core. Paul Barrett was a graduate student at the time but remembers how exciting it was when word of the discovery started to spread. 'It was the final clinching argument that birds really were dinosaurs', he says. 'It's a spectacular fossil, too. So it got a massive amount of attention and it was one of those moments that marked a step change in how things are done in a field. It had a massive impact at the time and immediately got a big backlash from people who didn't like the bird–dinosaur link.'

Since then, it's become increasingly difficult to say what unique attributes birds have and what defines them as a group. Writing in the journal *Nature*, anatomist and dinosaur aficionado Lawrence Witmer, of Ohio University in the United States, said: 'the discovery of very bird-like feathers ... in some of the predatory theropod dinosaurs found in Liaoning ... rocked the scientific world, because the feathered dinosaurs were outside the evolutionary group of acknowledged birds'.

The discovery of *Sinosauropteryx* was a big, big surprise to everybody, and 'it really changed everything in so many ways', Currie says. Only once in living memory had there been a similar amount of publicity about dinosaurs, when nests of dinosaur eggs were discovered in the late 1970s by Jack Horner and Bob Makela. 'People just went crazy because somehow finding eggs makes them seem more real. Somehow finding feathers and relationships to birds also makes them seem more real. It also makes you realise that they're not really extinct.'

Pack hunters

Phil Currie has been hooked on dinosaurs since the 1950s, when at the age of six he ripped open a packet of cereal and a plastic *Dimetrodon* tumbled out onto the table in front of him. By 11 he'd decided to become a palaeontologist – and not just anywhere, but in Alberta, because that's where he considered Canada's best fossils to be. In the intervening years Currie has contributed more to our knowledge of Canadian dinosaurs than perhaps any other person. He's been lauded with many medals, awards and honours, and most recently has had a museum named after him.

One of Currie's most controversial – and most interesting – ideas is that large carnivores, such as *Tyrannosaurus rex*, were not lone scavengers as has long been supposed, but pack hunters. While there was fossil evidence that smaller dromaeosaurs, such as *Velociraptor* and *Deinonychus*, hunted collaboratively, there was

no such evidence for larger species, such as allosaurs or tyrannosaurs. Few experts agreed with his idea, but Currie was convinced
that, at least some of the time, these predators had to collaborate
to get a meal.

If you've got 4-tonne herbivorous dinosaurs moving in herds
of hundreds or thousands of animals, they can't stay in one area
as their food requirements are too great, he argues; they must be
moving through many areas as part of a migration pattern. 'Then
you take it to the next step, and when you look at modern environments, which have big herds of plant-eating mammals, part
of the reason that these animals travel in groups is so they can
protect themselves while they're on the move', he says. 'And the
only way that carnivores can really respond to that is if they get
together in groups as well and pack hunt, so they can break up
these groups of plant eaters and chase out the weak and old.'

This pattern is something you see again and again in the natural world today – wherever there are open environments and big
herds of plant eaters traversing them, there are pack-hunting carnivores as well: wild dogs, hyenas and lions in Africa, for example,
and wolves in North America. Currie was convinced the same
kind of relationship would have existed between herds of herbivorous dinosaurs and the large carnivores that stalked them, but he
thought it was unlikely he'd ever find evidence for this because
carnivore fossils are so rare.

But it turned out that back in 1910, the AMNH's Barnum
Brown was working in a remote part of Alberta and found a site
where there were nine skeletons of a tyrannosaur called *Albertosaurus* all in the same quarry. Unfortunately, Brown had a small team
with him and they could collect only a small amount of material,
leaving the vast majority of the bones in place.

'There's a curious tie-in here', Currie says, 'because in 1996,
when I was in New York City [at the AMNH], and the *Sinosauropteryx* story broke, it was also when we found Barnum Brown's
collection. We realised he had collected parts of nine articulated

dinosaur skeletons from Alberta, which is a very exciting thing'. That was all the prompting he needed, and in 1997 he headed into Dry Island Buffalo Jump Provincial Park with only a few photographs and some scant notes taken by Brown in 1910 to relocate the site. 'Looking for dinosaurs is like looking for a needle in a haystack under the best of circumstances. But to re-find something on the basis of two photographs taken 90 years earlier doesn't work to your favour', he says, laughing.

But his team did rediscover the site and excavated the remains. 'The evidence is pretty overwhelming that this was a group of *Albertosaurus* that included everything from two-year-olds to 24-year-olds, which were all moving together', Currie says. Furthermore, it didn't appear that they were only scavengers that had gathered at the site to eat an already existing carcass. 'The inescapable conclusion is that this was a pack of some kind and I think Barnum Brown realised that back in 1910. He collected the material, but he ended up with so much material to describe from Alberta that he just never got around to doing anything about the fossils he collected from this site.'

The idea of pack-hunting tyrannosaurs is still hotly disputed, and others argue that the dinosaur remains may have collected together because of some kind of catastrophe such as a flood. Looking at the weight of evidence, though, it seems a simpler explanation that these animals were together in life and died together, but only time will tell. Currie and his colleagues have gone on to find three *Tyrannosaurus* skeletons together in a site in South Dakota in the United States, as well as handfuls of specimens in the same quarries for both *Tarbosaurus* (another *T. rex* relative) in Mongolia and *Mapusaurus* (a more distantly related carcharodontosaur predator that was even larger than *T. rex*) in Argentina.

As if the idea of a single *Tyrannosaurus* running you down wasn't terrifying enough, if Currie has his way, our nightmares might have to accommodate entire gangs of them. If nothing else, it would make for a fantastic movie plot.

Clinching the dinosaur–bird link

Initially, neither Currie nor any of the scientists in Beijing had been aware that Li Yinfang had sold the other half of the *Sino-sauropteryx* fossil to the IGP in Nanjing. Nor did the scientists in Nanjing know there was another half in Beijing. After seeing all the publicity around the Beijing specimen, Chen Peiji at the IGP had contacted Currie, who met him in the United States to look at photos of the specimen before deciding whether to come back to China to look at it.

'When Chen told me that he had the other half of the specimen, this was enough to floor me because although I knew the specimen was split, I didn't know there was another half', Currie says. 'I think I was probably the first person to see both sides of the specimen, other than the farmer who collected it in the first place.'

After the New York meeting, Currie returned to Nanjing with Chen and, in addition to the other half of the first fossil, was shown a second specimen of *Sinosauropteryx*, which also had feather impressions, proving the dark marks on the first fossil weren't any kind of accident of preservation. The feathers around that fossil had been pretty badly damaged by overzealous chiselling on the part of the farmers who had discovered and prepared it, but it was still clear what they were.

Meanwhile, in the United States, the debate had reignited between those who supported the idea that birds were descended from dinosaurs and those who were still not convinced. 'The arguments got even more intense', says Currie. 'They are feathers. They aren't feathers. They can't be feathers. They must be something else. All kinds of silly explanations came out.' One key argument was that the marks on the fossil were collagen fibres, perhaps from the muscles, or some sort of crest, but two later studies would prove that this wasn't the case on two counts. The first was the discovery in 1999 of beta-keratin (the keratin found in bird feathers) in the feather structures in the fossil of a feathered dinosaur called *Shuvuuia*. The second was a more recent electron microscope study showing the presence of melanosomes, tiny packets of pigment that give hair and feathers their colour, in the fluffy fringe of *Sinosauropteryx*. That study further suggested the fluff was ginger in hue and that these little dinosaurs may have had white stripes around their tails, something like ring-tailed lemurs. As Paul Barrett puts it: 'They look like feathers, they are made of the same stuff as feathers, and they are probably feathers.'

Whether or not the fuzz was made of feathers soon became a moot point in any case. 'My Chinese colleagues are industrious and don't wait around quietly while others are arguing over their specimens', Currie says. 'They went out and found more evidence.' Namely, the first dinosaurs with complex branched and vaned feathers of the kind found on modern birds. Ji Qiang and his colleague Ji Shuan discovered in quick succession *Protarchaeopteryx* and *Caudipteryx*, specimens of which Currie saw on his return to the National Geological Museum in Beijing in 1997. Like *Sinosauropteryx*, these fossils were found in Liaoning by amateur diggers who sold them on to dealers.

Protarchaeopteryx ('first ancient wing') has some similarities to *Archaeopteryx* (though not as many as its name suggests), but is more reptilian, despite appearing 15 million years later in the fossil record. It was a turkey-sized species without the asymmetrical feathers or body shape necessary for powered flight. *Caudipteryx* ('tail feather') was one of the parrot-beaked oviraptorosaurs. As birds have today, it had a short stiff section at the end of the tail, known as a pygostyle, but with a plume of tail feathers it probably fanned out for display purposes. It had downy feathers across its body and long feathers on its arms, although it, too, lacked the right kind of wings for flight. The species was described in *Nature* in 1998 by authors including Phil Currie, Ji Qiang, Ji Shuan and Mark Norell.

'The specimen was basically unequivocal', Currie says. 'There was no question it had feathers that were like birds'. They weren't just these down-like tufts that are on *Sinosauropteryx*. And it also had a skeleton that was perfectly intermediate between dinosaurs and birds.' One thing in particular excited him, and that was the ankle. 'Dinosaurs and birds unquestionably have very, very similar ankles. In fact, nobody has an ankle like a dinosaur or a bird. But there was some evidence embryologically in birds that suggested that they may have developed in two different ways', he explains. The shape of the ankle on both the new specimens was intermediate between the two types and provided very strong evidence that the ankle shape of birds had first developed in the dinosaurs.

The palaeo-ornithologists who had argued against *Sinosauropteryx* having feathers now changed tack. They agreed that the feathers in the new fossils were real – but first suggested that the *Protarchaeopteryx* specimen was a small dinosaur that had somehow died on top of a bird, then that neither *Protarchaeopteryx* nor *Caudipteryx* were dinosaurs at all, but were in fact secondarily flightless species that had evolved from bird ancestors that could fly.

But this makes no logical sense, Currie says. 'If you wanted to argue that *Caudipteryx* was in fact a bird because it had feathers then suddenly you've got a bird with the right kind of ancestral [ankle] condition for birds from dinosaurs. So it didn't matter whether you classified it as a bird or a dinosaur, there was that character, and many other characters as well, that weren't supposed to be in birds but were unquestionably in this animal. So it was perfect. I kind of considered it a trump card.'

Since then, nearly 40 species of dinosaur (see 'An A–Z of feathered dinosaurs'), and hundreds of specimens, have been found with evidence of feathers, and the number of experts who dispute the dinosaur–bird link has dropped to what Paul Barrett describes as 'a very small, albeit fairly vociferous minority'.

The palaeontologists who don't believe it, he says, provide no viable alternative for the origin of birds. 'They argue that birds come from some as-yet-unknown Triassic reptile', he says. 'But they can't provide any evidence that links birds more closely with any other group than theropod dinosaurs.'

Alan Feduccia, an emeritus professor at the University of North Carolina in the United States, is a highly respected authority on the evolution of birds, but on this point he disagrees with what has become the orthodox view for the vast majority of palaeontologists. 'I've studied bird skulls for thirty years and I don't see any similarity whatsoever to dinosaur skulls; I just don't see it', he told Pat Shipman, the author of *Taking Wing: Archaeopteryx and the evolution of bird flight*. 'I'm of the opinion that the whole hot-blooded dinosaur thing is very seriously flawed.'

Paul Barrett says he believes there is cultural dislike of the idea of birds as dinosaurs among a number of ornithologists. 'Some ornithologists have a real ingrained dislike for the idea of birds as dinosaurs – they are too vulgar. They'd much rather that the origin of birds was a holy mystery than that they are actually related to dinosaurs.'

In contrast, most people who work on fossil birds would agree they are a subset of dinosaurs. There remains a very small hard core of experts who don't like it. 'The scientific arguments they propose are weak and very easily demolished by evidence', Barrett says. 'It's just denial. They cherrypick evidence to support presupposed ideas they have.'

The vast weight of evidence now suggests that birds are really just a particularly successful experiment in being a small meat-eating dinosaur. In fact, as there are 10 000 of them, birds are the most successful dinosaurs of all, and were the only ones to survive the mass extinction at the end of the Cretaceous period. 'They are very evolutionarily successful, but on a very narrow body plan', says Barrett. 'Dinosaurs are now more successful than they've ever been, but they all look the same. With the exception of a few aberrations, they are all small, bipedal flyers. If you strip them down to the skeletons, there's not a vast amount of difference between them.'

The late 1990s were the end of the beginning for the story of China's feathered dinosaurs. It was time for a new player to enter the game – a brilliant young Chinese man who would discover and describe more dinosaurs than anyone in living memory. A revolution in dinosaur science had begun.

3

The dinosaur hunters

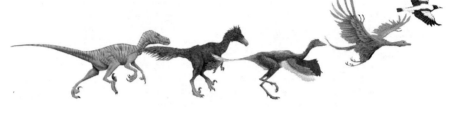

China has hit the fossil mother lode, and a staggering number
of new dinosaurs has been unearthed there in the last 15 years.
The boom calls to mind an earlier, less harmonious burst of
dinosaur discovery.

In April 2005, Xu Xing was working with a Japanese documen-
tary film crew to re-enact his earlier discovery of a long-necked
sauropod called *Sonidosaurus* at a dig site in Inner Mongolia's Gobi
Desert. Crouched in the dirt on the bank of a dry river in the
Erlian Basin, Xu and Lin Tan, his local collaborator, selected a
large fragment of thighbone at random and began to brush the
dirt away to demonstrate the art of fossil collection. But the film
crew got more than they'd bargained for, ending up with footage
of a much more candid nature. Xu, the world's foremost expert on
feathered dinosaurs, became visibly excited as he worked, real-
ising the leg bone wasn't from a sauropod at all. It was from an
unknown theropod – and something big – potentially as big as
T. rex. 'I told them to stop filming', Xu later recalled. 'I said, "This
is not for your program".'

Further digging revealed a hind limb, a forelimb, a pelvis,

a lower jaw and vertebrae. The species he'd found was a massive parrot-beaked oviraptorosaur that would come to be known as *Gigantoraptor erlianensis*. Dinosaurs in this family – such as *Oviraptor* and *Caudipteryx* – are typically no bigger than an emu, and would have been no heavier than 40 kilograms. With a length of 8 metres and weighing more than 1400 kilograms, *Gigantoraptor* was the size of a bus and could have looked *T. rex* in the eye. The fossil didn't reveal feather marks, but as other oviraptorosaurs are known to have been feathered, it's very likely that *Gigantoraptor* was too, making it one of the largest known feathered animals in history. Rather than for insulation, the plumage was probably used for mating displays and shading clutches of eggs in the heat of the sun 85 million years ago.

The fact that Xu had made a scientific discovery of such magnitude more or less by accident typifies the kind of remarkable good fortune that has been a trademark of his career. A groundbreaking young professor who's had his own share of adventures scouring Asia for priceless prehistoric artefacts, he has been labelled China's answer to Indiana Jones. There's some truth to this – still in his early 40s he has already been involved in the discovery and naming of more than 50 dinosaurs, including many feathered species, and he has a passion for journeying into the unknown.

This impressive tally of new species places Xu among a few scientists who stand out in history for discovering and describing more dinosaurs than anyone else. Others include Yale University's OC Marsh (1831–99), and two living scientists, Dong Zhiming of China and Jose Bonaparte of Argentina, both of whom are now elderly. '[Xu] has described more new kinds of dinosaur than any person who has ever lived', says Peter Dodson, a palaeontologist at the University of Pennsylvania in Philadelphia. '[He] is surely the new kid on the block ... [and] absolutely the right person at the right time.'

Phil Currie agrees that Xu was the right person at the right time, and represented a new generation of scientific talent in

China. 'A lot of the people who worked in China up to that point had gone through the Cultural Revolution', he says. 'They were certainly having a hard time retraining themselves and catching up with everything that had been happening in the West. But Xu, being a young person, was more flexible in terms of developing his ideas, and also assimilating the ideas of other people from around the world. He's also such a nice person, so everybody liked him and wanted to work with him.'

Building up a rich network of contacts in the West – as well as a fair bit of fame for his discoveries of jaw-dropping feathered dinosaurs – has helped Xu reach heights of international academic success rarely achieved before by Chinese scientists.

Xu Xing: student prodigy

When Xu was at school, he says, he barely even knew what a dinosaur was. All that really interested him were physics and mathematics, and he hoped to become a physicist. But this isn't how the university system works in China, and he wasn't free to study whatever he wanted as many of us are in the West.

'I came from Xinjiang, a very remote area in China, so education there wasn't so good', Xu says. 'When I was a kid I don't remember knowing anything about dinosaurs or even palaeontology. I was assigned [by the government] to the major of palaeontology at Beijing University, so it was a kind of an accident for me to go to this university and do palaeontology.'

He describes Xinjiang, a Muslim-majority province bordering Kazakhstan, as isolated and backwards at that time. His parents moved there as part of a Cultural Revolution initiative in the 1960s that saw educated people relocated to rural areas.

In the last year of his master's degree he had to prepare and study dinosaur fossils for his thesis. Some of these fossils turned out to be the oldest ceratopsians (relatives of *Triceratops*) ever discovered, and they extended the history of the group back from the Cretaceous to the Late Jurassic. 'He demonstrated that ceratopsid dinosaurs inhabited eastern China', says Peter Dodson. 'Previously, ceratopsids, the great horned dinosaurs, were known only from western North America.'

It was during this early work on dinosaurs that Xu gave the topic real consideration for the first time. 'It was then I realised that palaeontology was something maybe interesting', he says. 'I started to have some interest in dinosaur palaeontology especially. Then after I got my master's degree, I got a job at the [Institute of Vertebrate Paleontology (IVPP) in Beijing] ... I was lucky in my first few years because I got a chance to study some really interesting fossils. That made me really addicted to this field.'

Xu may have been a slow starter, but he's more than made up for lost time since then. All up he's described around 60 species of fossil animals. His most significant finds have firmed up the link between dinosaurs and birds, and started to flesh out all the gaps in between. He's a workaholic who says he believes he was destined to study dinosaurs and now can't imagine life without them.

The discovery of the first feathered dinosaur, *Sinosauropteryx*, in 1996, was quickly followed by *Protarchaeopteryx* and *Caudipteryx*. All three were described by researchers led by Ji Qiang, who is now at the Chinese Academy of Geological Sciences in Beijing. These remarkable discoveries marked Liaoning, and in particular the Yixian Formation, as one of the most astonishing sources of dinosaur fossils of all time.

The discoveries by Xu's team are mostly small carnivorous dinosaurs from Liaoning, but also include other types of dinosaur, such as ornithopods and ceratopsians, and his fossils have also come from Shandong, Inner Mongolia, Gansu, Yunnan and Xinjiang. They described their first dinosaurs in 1999 – *Beipiaosaurus* and *Sinornithosaurus* – and a steady stream of descriptions has followed ever since. 'Collectively these dinosaurs have revolutionised our understanding of dinosaurs and their relationship to birds', says Dodson. 'The [family tree] and the origin of feathers are both clearer now than they ever have been.' Not only did these fossils settle decades of debate, proving that birds were the living descendants of the great dinosaurian dynasty, but they also started to fill in the many details in between of how one such seemingly different animal evolved from the other.

The Chinese fossil rush

Peppered with farmland and factories, bordering North Korea and the Yellow Sea, Liaoning Province in north-eastern China is a day's drive from Beijing. It has been a hub of palaeontological activity since the early 1990s. The feathered fossil of *Sinosauropteryx* spurred a fossil-hunting gold rush there like no other. Since then, Liaoning has become famous for producing fossils that vividly reveal entire ecosystems of species from the Early Cretaceous.

The rocks of several formations in Liaoning, including the Yixian, are some of the richest in the world for yielding fossils, and have dramatically changed our understanding of life during the Late Jurassic and Early Cretaceous eras. Formations are layers of rock laid down, without breaks to divide them, over a relatively short period of geological time; the Yixian Formation was deposited from 121 to 125 million years ago at the beginning of the Cretaceous. Fossils from Liaoning as a whole span the entire period from 120 to 160 million years ago.

Fossils typically form when the remains of plants and animals are buried by sediments carried by water. How long it takes for burial to occur makes a big difference to the quality of the fossil – those left in the open for longer are more likely to be eaten, become heavily decomposed and also have their constituent parts separated. Most of the fossils of Liaoning are special because volcanic ash rapidly buried them after death or was responsible for the death in the first place.

Many are preserved as flattened, largely two-dimensional specimens (something akin to prehistoric road kill) in sheets of shale made of fine-grained sediments laid down on lake bottoms. But others were trapped in coarse-grained sediments from river bottoms, which preserved them more three-dimensionally. In those sediments, animals have even been found frozen in poses suggestive of prehistoric behaviour, perhaps because they were rapidly swallowed by flows of mud and sand during volcanic eruptions or other cataclysms.

At the time the fossils formed, Asia consisted of a series of isolated basins that favoured the formation of lakes and marshy habitats. Around these lakes grew lush forests of non-flowering plants, such as ginkgoes, conifers, cycads and seed ferns. Closer to the ground were horsetails, lycopods, ferns and early relatives of today's flowering plants. The whole region was dotted with volcanoes – and eruptions, bushfires and upwellings of noxious gases from the lake bottoms would have periodically annihilated entire ecosystems and helped create a wide variety of dynamic habitats, which accounts for the great diversity of plants and animals found in Liaoning at this time.

Vigorous volcanic activity ensured that a rain of ash frequently smothered the lakes, creating fine-grained sediments that were starved of oxygen and perfect for preserving the animals silently suffocated by the noxious volcanic gases. '[These] lake beds combined with some volcanic input appear to be particularly good at generating exceptional preservation', says the University of Montana's David Varricchio. A great diversity of species is preserved in the shales these sediments produced, often in exquisite detail, right down to the gut contents, feathers and scales – and not only dinosaurs and birds, but also flowers, crustaceans, insects (including the delicate patterns on their wings), fish and early mammals.

Liaoning is now offering such great numbers of fossils that Chinese scientists are having trouble studying and naming them all. 'In my office, in my colleagues' offices, there are many, many new fossil specimens', Xu says. 'I have quite a few new species waiting to be described.'

Lucky break

Xu made his first trip to western Liaoning in 1997. He and his colleagues were aware of great fossils coming out of the region and were hopeful it could yield more spectacularly preserved dinosaurs.

'That year we organised a small team and we went to find more feathered dinosaurs', he says. Several weeks of digging near the city of Beipiao turned up many birds and other fossils but no dinosaurs. They were almost ready to pack up for that year, but on the last night of the field season a farmer came and told them he had collected some fragmentary fossils several years earlier. 'He said if we wanted to research them, he would donate the fossils to the institute', Xu says, 'so we were very excited ... And then he gave us those very, very fragmentary specimens. Hundreds of small pieces. Finally we collected the pieces into a relatively large block, and they turned out to be ... a *Beipiaosaurus*, a new feathered dinosaur'.

Beipiaosaurus inexpectus ('unexpected lizard from Beipiao') strongly suggested that feathers had probably been widespread across the dinosaur group. *Beipiaosaurus* was a shaggy and weird therizinosaur. Therizinosaurs were a group of carnivorous theropods whose teeth suggest they had returned to a vegetarian diet. Puzzlingly, *Beipiaosaurus* also had long arms with large scythe-like claws, which may mean it led a similar life to the extinct giant ground sloths of South America, which used their claws to pull down vegetation from the trees. Therizinosaur means 'reaping lizard or scything lizard'.

From then on, Xu's IVPP team organised many expeditions and excavations in Liaoning and neighbouring areas in Inner Mongolia and Hebei Province. The feathered dinosaur fossils kept rolling in – *Sinornithosaurus, Pedopenna, Microraptor, Dilong, Mei* – and with them important insights into the origins of feathers, flight, bird-like brooding behaviours and much more. 'We were very, very lucky', Xu says. 'We found many, many new feathered dinosaurs.' Today his lab at the IVPP has an academic staff of around 15 – along with seven fossil preparators (a number almost unheard of elsewhere) and five postgraduate students – to help him shoulder a vast workload.

When Xu started his career he says he had trouble finding even a single fossil and could never have imagined the success he has enjoyed since. 'Every time we made another discovery ... we thought maybe this is the last major discovery I will make in my whole career. Any one of these would have been a major discovery for one person ... Normally if you make a discovery like this, then you'd be happy for the whole of your career. But every year we'd make another discovery.'

Xu's success is down to skill, luck and tenacity, but also to how common fossils are in Liaoning. There is a surfeit of farmers looking for them to sell on to fossil dealers and museums – an illegal practice that has led to a problematic black market in specimens, some of which are fake or have been manipulated (see chapter 5). Increasing urban development and construction in the 1990s was another reason fossils started to be found more readily. 'In effect, [Xu] has an army of collectors working for him in Liaoning – collectors who are farmers with a cottage industry of finding fossils', says Peter Dodson. 'Many of these fossils come to Beijing, where he eagerly describes them. He has also conducted a highly successful collecting program in Inner Mongolia and Xinjiang in remote north-western China.'

The fossils are everywhere, even in farmers' own backyards, so locals can make money from collecting them, Xu says. 'Although it is illegal for local farmers to collect fossils, western Liaoning and the neighbouring areas are very poor. People want to make a better life for themselves, so they decide to collect fossils and sell them to make money.' Xu is skilled at bargaining and negotiating, and has set up a network of contacts to help ensure that the most important specimens come into the hands of institutes and museums rather than being lost forever to private collections.

The 'bone wars'

Only once before in history has there been a dinosaur-hunting gold rush anything like the kind now happening in China. The fossil hunters at the centre of it were certainly as productive in describing species as Xu, if far less likeable. It's a fascinating story that's little known outside the United States or by those unfamiliar with palaeontology. Yet it's a great historic tale full of animosity and intrigue, a tale of the untamed American West and of two rivals – Edward Drinker Cope and Othniel Charles Marsh – who hated each other with such passion that it would ultimately ruin them. But from the spoils of this great battle – or the 'bone wars' as the newspapers of the late 1800s labelled it – would come a completely new picture of the history of life on earth.

The term 'dinosaur' and the group Dinosauria were only coined by Richard Owen, director of natural history at the British Museum, in April 1842. He intended them to mean 'fearfully great, a lizard', though ever since they have been poorly translated in most instances as 'terrible lizard'. Many of the earliest discoveries of dinosaur fossils were in England, although work in the 1850s by Joseph Leidy in Haddonfield, New Jersey, led to the discovery of *Hadrosaurus* and the evidence that many dinosaurs were bipedal, rather than the sprawling quadrupeds Owen had imagined.

Before 1870, only nine species were known from North America, but by the late 1890s ED Cope and OC Marsh had between them described 140-odd new species. They were responsible for discovering and naming many of the famous dinosaurs most people are familiar with today, including *Allosaurus*, *Triceratops*, *Stegosaurus*, *Apatosaurus* and *Diplodocus*. These were the largest animals ever to walk across the face of the earth – and they were far, far larger than anyone could ever have imagined.

Marsh, the nephew of wealthy industrialist and philanthropist George Peabody, was Yale University's – and the world's –

first professor of palaeontology, while Cope was a precocious teen who became professor of zoology at Haverford College in Pennsylvania, and went on to several other academic appointments, including a position at Philadelphia's Academy of Natural Sciences. Cope was described as 'pugnacious and quarrelsome' by one of his contemporaries, William Berryman Scott, but he published more than 1400 papers over his life (the first, on salamanders, at the age of 18), making him the most prolific author in the history of scientific enquiry. He also described more than 1000 fossil species, including hundreds of fish. Marsh – himself described as autocratic, petty and miserly – was a more methodical worker than Cope, and less prone to rush into print with new discoveries, but he also described more than 500 species.

Cope and Marsh spent 30 years scouring the newly opened up American West for fossils and frequently coming into conflict with one another. They are credited with laying the foundations of modern palaeontology and creating a public passion for enigmatic prehistoric animals that continues to this day.

In the wake of Darwin

The cause of this great awakening of interest in palaeontology and fossil species in the latter part of the 19th century was the publication in 1859 of Darwin's *On the Origin of Species*. The idea of evolution and life changing over time was well established at this point, but Darwin's theory of natural selection was whipping up a storm of controversy.

Many leading American thinkers of the day found Darwin's theory difficult to swallow, but Marsh wasn't one of them. He worked methodically to collect fossils that would reveal the evolutionary history of groups of animals, and he believed that the arid and open American West would surely yield the fossils that would prove Darwin right. 'I felt that entombed in the sandy clays ... there must be hidden the remains of many strange animals new to science, long waiting to be brought to life', Marsh later recounted.

Perhaps the most remarkable was his discovery of 33 species of fossil horse, which illustrated the branched history of the group from fox-sized, three-toed creatures of the early Eocene (52 million years ago), to the imposing, single-toed equids we're familiar with today. Url Lanham, author of *The Bone Hunters: The heroic age of paleontology in the American West*, describes the horse fossils as 'by far the longest and most continuous series known for an actively evolving group of organisms. It provides the kind of direct, incontrovertible evidence that makes of evolution a fact rather than a theory'. Marsh's studies of the fossils of early toothed birds – including *Ichthyornis* and *Hesperornis* – published in 1880, also helped confirm their evolution from reptilian ancestors.

These finds impressed both Darwin and his 'bulldog', Thomas Henry Huxley, across the Atlantic, and Marsh met with them on several occasions. Huxley was particularly blown away by the fossil evidence of the horses when he visited Marsh at Yale in 1876. On 31 August 1880, after receiving Marsh's tome on fossil birds, Darwin wrote to him, saying:

> I received some time ago your very kind note of July 28th, and yesterday the magnificent volume. I have looked with renewed admiration at the plates, and will soon read the text. Your work on these old birds and on the many fossil animals of N. America has afforded the best support to the theory of evolution which has appeared within the last 20 years.

While Cope, a Quaker with deep religious convictions, believed that animals evolved over time, he did not support Darwin's idea of natural selection. As historian James Penick noted, 'Cope saw in one reading of the *Origin* that it endangered his faith, and he wrote dozens of essays in the years that followed to exorcise the threat'.

Marsh and Cope's difference of opinion over the mechanism of evolution was only one of many sources of enmity between them. In 1863, when they first met in Germany, at Berlin University, they were on friendly terms, even naming species in honour of each other upon their return to the United States. Things would begin to turn sour in 1868 after Cope showed Marsh around the fossil-rich quarry site in Haddonfield, where his mentor Joseph Leidy had discovered *Hadrosaurus*. Cope's fossil supply from there dried up thereafter, and it transpired that Marsh had made an underhand agreement with the quarry owner to have further fossils sent to him at Yale's Peabody Museum. The hatred between the two would be sealed shortly afterwards, when upon a visit to the Academy of Natural Sciences in Philadelphia, Marsh pointed out an embarrassing mistake Cope had made in reconstructing a plesiosaur called *Elasmosaurus*. Cope had accidentally put the backbone the wrong way around, affixing the head to the tip of the tail. He desperately attempted to buy all the copies of the learned journal in which he had described the animal, but it was too late to hide his mistake. Marsh later wrote, 'when I informed Professor Cope of it, his wounded vanity received a shock from which it has never recovered, and he has since been my bitter enemy'.

In subsequent years, as both they and their teams of diggers and preparators scoured the bone beds of Utah, Colorado, Wyoming, Nebraska and Kansas, the pair exchanged heated letters, and later threw insults across the pages of journals and newspapers. One 1873 letter Marsh wrote to Cope after discovering he had poached one of his field workers and stolen his fossils read: 'The information I received on this subject made me very angry and had it come at the time I was so mad at you for getting away Smith [the poached field worker] I should have "gone for you," not with pistols or fists, but in print ... I was never so angry in my life.'

Marsh paid spies to keep a watch on Cope and exchanged coded telegraphs with them, referring to Cope by the pseudonym

'Jones'. His fossil collectors would smash up bones if they had more than they could ship away from dig sites, just to prevent Cope's workers from getting them. Both Cope and Marsh frequently published papers describing species the other had already labelled and then squabbled over the naming rights.

'As their intense competition to uncover dinosaur bones raged across the fossil fields of the American West', wrote Tom Huntington in *American History* magazine:

> Cope and Marsh quarrelled continuously in the press and amid the government circles of the nation's capital. As a result, not all of the animals that they described became permanent additions to the roster of extinct species.
>
> Their race for pre-eminence sometimes caused the two palaeontologists to give different names to the same species and announce discoveries of new animals without having adequate evidence. Yet while their mutual hatred often expressed itself in petty ways, it did spur activity in the field and greatly increased man's knowledge of extinct creatures.

Tripping over new species

And the remarkable fossils kept pouring in. In 1870 Marsh was heading an expedition with students from Yale, which briefly saw soldier and frontier hunter Buffalo Bill Cody guiding them through a stark landscape under threat of attack from hostile Native Americans. In a narrow canyon in western Kansas, Marsh found fragments of what appeared to be a species of giant flying reptile – it was the first pterodactyl discovered in the Americas and it was an enormous one at that. Marsh described it as 'truly a gigantic dragon even in this country of big things, where hitherto no Pterodactyl large or small had yet been discovered'. Measurements of the bones indicated 'an expanse of wings of not less than twenty feet!' he wrote in a paper describing the find. He named the species *Pterodactylus oweni*, 'in honor of Professor Richard Owen, of London'.

Marsh's assistant, John Bell Hatcher, was in Wyoming in the late 1880s when local cowboys led him to a great skull they had broken free of the wall of a canyon. It had 'horns as long as a hoe handle and eye holes as big as your hat', Hatcher wrote. The horns had been broken off as it had fallen to the canyon floor, but he retrieved it and presented it to Marsh, who eventually named the creature *Triceratops*.

Up to this time, a number of competing geological survey departments had been working under the banner of the US federal government, and they had variously been funding either Cope or Marsh's fossil hunting expeditions, but in 1879 Congress decreed that they be consolidated into a single organisation: the US Geological Survey (USGS), headed by John Wesley Powell, a well-respected explorer of the West. Powell appointed Marsh as the nation's 'chief palaeontologist'. This couldn't have been a worse turn of events for Cope, whose government funding dried up almost overnight. His own personal fortunes were dwindling, and his investments in a series of disastrous silver-mining ventures over the next decade meant he was heading towards bankruptcy by the late 1880s. He had one resource left, though, that he could sell to help keep him afloat – his vast collection of fossils, which filled the Philadelphia apartment he had been reduced to living in.

But Marsh conspired to rob him even of this. In 1889 he pointed Powell to a clause in the USGS charter that said all fossils collected with government funds should be turned over to the Smithsonian Institution in Washington, DC (Marsh had of course negotiated that he would be able to hang onto his own collection). Marsh then went to the secretary of the Smithsonian and suggested they take Cope's fossils. Cope, who had detailed records going back decades, showed that many of his specimens had been collected with more than $80 000 of his own money rather than with government funds and won the argument. He was eventually able to sell part of his collection (13 000 specimens), amassed over 20 years, to New York's AMNH for $32 000.

Now Cope was out for revenge. He collected together 20 years' worth of jottings and clippings detailing Marsh's faults and errors, which he called his 'Marshiana', and went to a reporter he knew at the *New York Herald*, the most widely read American newspaper of its day. On Sunday 12 January 1890, a headline appeared that read: 'SCIENTISTS WAGE BITTER WARFARE – Prof. Cope of the University of Pennsylvania brings serious charges against director Powell and Prof. Marsh of the geological survey.' In the article, Cope accused Marsh of fraud, plagiarism and incompetence, saying he paid people to do his work but stole all the glory for himself. Marsh's body of work, he said, was 'the most remarkable collection of errors and ignorance of anatomy and the literature on the subject ever displayed'.

The print war raged back and forth for three weeks with damaging accusations artfully volleyed from both camps. Marsh saved his most damning accusation for last, telling the tale of how Cope had affixed the head of the *Elasmosaurus* fossil to the tip of its tail.

The dust settled, but the scrap had left a bitter taste in the mouths of the American establishment. In 1892 Congress began to question the power and expenditure of the USGS. One US congressman cited Marsh's work on the toothed fossil birds as a particularly pointless area of expenditure. The fact that Darwin had called this the greatest proof for the theory of evolution was all but forgotten, and the catchphrase 'birds with teeth' became synonymous with government waste.

The budget of the USGS was slashed, Marsh had to give up his position and – indignity of indignities – he himself was instructed to return 80 tons of his own fossils to Washington. Less meticulous than Cope, Marsh had little documentation and had no choice but to give them up. For the first time in his life he had to draw a salary from Yale. In 1897, aged 56, Cope died of kidney failure. Marsh died two years later, aged 67, from pneumonia.

Dinosaurs roam once more

Though much of the work of these feuding fossil hunters was little appreciated by the public during their lives, dinosaurs would soon start to loom very large in people's minds. The excitement and controversy of the scandal is perhaps what first propelled dinosaurs into the public consciousness. Before Cope and Marsh few people had paid much attention to what little was known about these long-gone giants, but within a decade of their deaths, titanic skeletons were being reconstructed in museums across the eastern United States. 'Big dinosaurs were going up in American museums almost as fast as skyscrapers were rising in New York or Model Ts were being turned out in Detroit', writes Mark Jaffe in *The Gilded Dinosaur*. He recounts how the Carnegie Institution found an enormous specimen of *Diplodocus* in the Badlands of Wyoming in July 1899, just a few months after Marsh died. The institution's founder, Scottish-born millionaire Andrew Carnegie, employed Marsh's former assistant John Bell Hatcher to reconstruct it, and when completed it was more than 26 metres long.

King Edward VII commented to Carnegie that one day he might like the British Museum to house such a specimen. That was all the prompting Carnegie needed – he had a cast of the specimen made at great expense, and after 18 months of toil it was shipped to London in 36 crates. 'Dippy' the dinosaur went on display at what is now known as the Natural History Museum in May 1905, and has since been enjoyed by many millions of visitors.

Dippy remains on display in the Central Hall today and has no doubt been inspirational to many thousands of small children. Intriguingly, casts of the same *Diplodocus* rapidly ended up in museums around the world. Kaiser Wilhelm II requested one too, and soon the Carnegie Museum was making copies that ended up in Berlin, Paris, Vienna, Bologna, St Petersburg, Madrid, Mexico City and even La Plata in Argentina.

Most sources agree that the feud between Cope and Marsh stimulated the search for fossils and interest in the prehistoric past, but much of it was of the 'smash-and-grab variety', with great piles of fossils remaining packed in the cases they'd been collected in, never opened and examined. Without the feud, although modern palaeontology would surely have blossomed later, many of the 130-odd dinosaurs described by the pair would still lie unknown within the arid plains, canyons and cliff faces of the great American West.

Great game hunters

While the prodigious record of discoveries by Xu Xing and his colleagues in China are generating a modern fossil-hunting gold rush that echoes that of Cope and Marsh, their method of enquiry is far more carefully considered. There is little doubt, however, that we owe a great part of our knowledge of the dinosaur era to these tenacious hunters of the great game of the Cretaceous and Jurassic.

'The so-called "bone wars" in North America, from 1870 to 1890, [were] a time of intense activity during which up to 15 species were named each year by arch rivals Edwin Cope and Othniel Marsh', writes British palaeontologist Mike Benton. 'After their deaths, species discovery waned, and there were particularly low levels of work through the two world wars. The time since 1990 has seen a second, even more extraordinary, phase of discovery and naming of new species, some 30 per year.'

Bursts of new discoveries occur as geological formations in parts of the world yet to be fully explored palaeontologically are targeted by fossil hunters – the same thing has occurred in Mongolia, Argentina and North Africa – but 'there will surely be fewer and fewer such unexplored new basins as time goes on', says Benton. The records of Cope, Marsh, Xu, Dong Zhiming, Jose Bonaparte and others among the top-10 dinosaur describers and hunters of all time may therefore never be surpassed.

4

From dinosaur to bird

From metabolism and genetics to diseases and behaviour,
many of the things we associate with birds first evolved
in the mighty dinosaurs.

Sometime in the early Cretaceous, a duck-sized dinosaur scratches, stretches and curls up among the fern fronds near the forest floor to sleep. She folds her hind limbs up beneath her body, and her feathered forearms in against her sides, before turning her long neck back and tucking it behind her elbow in much the same way birds roost, with their heads under their wings. The position is one this little meat eater adopts each day come nightfall, and it helps her keep warm in the chilly air. During her peaceful slumber a blanket of noxious gas from a volcanic eruption suffocates her, while ash rains down upon her, leaving her frozen in this sleeping posture and buried underground near Beipiao, Liaoning, for 130 million years.

That was until she was excavated and described in 2004 by Mark Norell, of the AMNH in New York, and Xu Xing, of the IVPP in Beijing. They called her *Mei long* – Chinese for 'soundly sleeping dragon'. Finding clues to behaviour in fossils is rare, but when that behaviour is sleep it represents an extraordinary discovery. Not only did *Mei* hint that many common behaviours of modern birds evolved in their dinosaur ancestors, but she also added to the evidence that dinosaurs were active, warm-blooded creatures, similar to birds and mammals.

'This specimen displays the earliest recorded occurrence of the stereotypical sleeping or resting behaviour found in living birds', Xu and Norell wrote in their *Nature* paper on the find. 'In birds, the tuck-in posture reduces surface area and conserves heat in the head, a major region of heat loss in these animals. It is therefore usually associated with heat conservation.' The fact the animal was a primitive member of the troodontid group of small carnivores also suggested that the tuck-in sleeping posture was adopted in dinosaurs long before they evolved into birds.

In January 2012 the discovery of a second specimen of *Mei long* was reported, and it was in precisely the same bird-like roosting position, reinforcing the idea that the fossils were capturing a typical behaviour rather than a freak incident. As British author Colin Tudge writes, 'It is pleasing to think that behaviour we assume is exclusively avian in fact was shared by some long-gone creature an almost incomprehensibly long time ago – not a monster at all, but rather sweet'.

Ancient heritage

The roosting behaviour of *Mei* is just one example of many, many traits and behaviours that were once thought to be the exclusive province of birds, but which we now know were shared with the dinosaurs. Of course these include such things as skeletal features, feathers, flight, beaks and nesting behaviour, but also more subtle similarities such as diseases and aspects of metabolism. Clues to many of these were found with new fossils; others came from existing dinosaur fossils in museum collections when experts re-examined them in light of the fact they were closely related to birds.

'If you took an evolutionary tree [of dinosaurs and birds], and marked off all the features that we used to think of as unique to birds, you'd see loads of them go all the way back down that family tree now', says Paul Barrett. 'They are spread out along that branch rather than all being clustered where birds appear. Lots of things we thought were unique to birds go way back down the theropod family tree.'

Breathing system

One of these traits we now know evolved long ago is the unusual and specialised breathing apparatus birds have. Unlike mammals (which breathe by contracting and expanding their lungs to take air in and out via the same route) birds have a one-way system to pump air right through their bodies. This means they are much more efficient at getting oxygen in and out of their blood than mammals, and – combined with their light bones – it's likely to have been a major factor in helping them take to the air, given flying requires great power and energy.

Birds have a small, rigid lung fixed to their backbone and attached to a massive system of flexible air sacs and capillaries. These work their way between the organs and tissues, and even invade the hollow bones, helping to create the honeycomb structure that makes them a fraction of the weight of mammal bones. This feature of bird bones is called pneumatisation.

'A bird lung is just a stop along the way for air that's flowing around this big system', Barrett says. 'The air goes in through the lungs, through this very complicated set of outpocketings of the lungs, and then out again. It's a one-way system and it occurs in the opposite direction to blood flow, so it's amazingly efficient.' The oxygen content of the air in the lungs is at its highest when it comes in, and this is where the oxygen content of the blood is at its lowest, which results in excellent oxygen exchange between the two systems. This means birds can sustain the very high metabolic rates essential for flight.

There was already evidence that dinosaurs had the same kind of respiratory set-up, but a fossil of an *Allosaurus* relative revealed in 2008 clinched it. The wishbone, hipbone and ribs of the 10-metre carnivore discovered on the banks of Argentina's Rio Colorado were filled with spaces and had the spongy look characteristic of pneumatisation. The experts behind the discovery – including David Varricchio of the University of Montana and Paul Sereno of Chicago's Field Museum – named the species *Aerosteon riocoloradensis*.

More than any other dinosaur discovered so far, this fossil provided telltale signs that dinosaurs shared the bellows-like breathing system of birds. There was also a tantalising clue that *Aerosteon* – as heavy as an elephant – might have used the air sacs as a cooling mechanism as well. Cooling can be a significant problem in large animals, which lose heat more slowly to the environment than small animals (particularly dinosaurs, which didn't have sweat glands as we and other mammals do). According to Sereno, *Aerosteon* had air sacs in unusual places, which came around the outside of the body and entered ribs in the belly region. It also appeared to have a network of tubes under its skin. Could it be that it was using them to dissipate body heat to the environment? Only more fossil evidence will tell, but it's an interesting idea.

Many other fossilised dinosaur bones have cavities in them suggestive of air sacs, which leave characteristic marks on bones. Most theropods and some sauropods have these holes, strongly hinting that they had bird-like one-way lungs. The fact that the feature is widespread across terrestrial dinosaurs – and their close relatives the flying pterosaurs – suggests it appeared before either group evolved.

But if it wasn't for flight, it's not completely clear what the purpose might have been. One idea is that it helped dinosaurs attain massive size by allowing them to get oxygen into body tissues that would otherwise have been too far away from their lungs. Another suggestion is just that dinosaurs and pterosaurs were very active animals with high metabolisms, and the efficient breathing system allowed this. 'People used to think the air sacs were simply associated with saving weight – and that makes a lot of sense with the giant size of dinosaurs', says Mike Benton of the University of Bristol. 'But many are now convinced that if they did have a sort of avian respiration with one-way flow of oxygen, that would have enhanced their efficiency.'

Other studies have shown that a number of dinosaurs, including *Velociraptor* and the 'first bird' *Archaeopteryx*, may have had bony levers (called uncinate processes) on their ribs. These work in birds to help move the ribs and sternum efficiently during breathing. The diaphragm of mammals allows them to force air into and out of the chest cavity. Instead, birds use muscles attached to both these rib-mover bones and their sternum to pump air through the sacs in their chest cavity effectively.

Modern birds occasionally use the air sac system for display purposes or to produce sounds (such as the male great frigatebird, found on tropical islands across the world, which inflates a large, red, balloon-like sac on its throat to woo females). Since dinosaurs had the same system of air sacs, it seems reasonable to assume they may have occasionally employed it for display purposes. While the thought of a dinosaur inflating a throat sac the size of a small car in an attempt to woo a mate is an intriguing one, of this the fossil record has sadly left no trace.

Growth and development

Development and growth is another area where dinosaurs are thought to have been surprisingly bird-like. Before the 1990s the prevailing wisdom was that dinosaurs grew in a similar way to living reptiles, such as crocodiles, which slowly increase in size throughout their lives, only attaining their greatest size in old age. But many studies of fossil dinosaur bones now hint that they, in a similar way to birds, underwent much more rapid bursts of growth in early life to reach a large size relatively swiftly.

'Whereas dinosaurs may not have grown at rates exactly like those of extant birds and mammals, they seem generally to have been more like them than like other extant reptiles', wrote Kevin Padian and other experts in *Nature* in 2001. When bones grow they create microscopic structural features that – somewhat similar to tree rings – can tell us about their growth rates. These features in the dinosaur bones showed a clear similarity to those in birds. The discovery was significant, because it offered clues as to how some dinosaurs, particularly the long-necked herbivorous sauropods, could grow larger than any other land creatures in the history of life on earth. The largest sauropods are now known to have weighed in at more than 100 000 kilograms, equal to about 20 African elephants.

In 2013 a remarkable cache of baby sauropod bones from Lufeng in China provided the best glimpse yet of how embryonic dinosaurs grew and developed. This collection of hundreds of minute fossil bones – many no wider than the lead in a pencil – and pieces of eggshell dates to more than 190 million years ago, in the Jurassic. As most juvenile dinosaur fossils are from the Late Cretaceous, this pushed the date of the oldest fossils of baby dinosaurs back by more than 100 million years. The bones are those of around 20 young *Lufengosaurus* (the most common dinosaur in this region in the Early Jurassic), and came from a number of different egg clutches at a range of stages of development. By comparing the length of the thighbones of young at different stages of development, experts led by Robert Reisz of the University of Toronto in Canada, showed they were growing at a faster rate than ever recorded in any other species of bird, mammal or dinosaur.

'We are opening a new window into the lives of dinosaurs', Reisz told reporters. 'This is the first time we've been able to track the growth of embryonic dinosaurs as they developed. Our findings will have a major impact on our understanding of the biology of these animals.' He argued that rapid, sustained growth of embryos, and brief incubation times in the egg, were factors that allowed *Lufengosaurus* to swiftly outgrow the carnivores that preyed upon them, and to reach lengths of about 9 metres.

More clues from the bones showed that the thighbones were being reshaped during growth, which suggested muscles were contracting and pulling against them. This was a clue that dinosaurs moved around inside their eggs, much as birds do, and was the first time this kind of avian behaviour had been detected in a dinosaur.

Warm-bloodedness

For several decades now the debate has continued to rage among dinosaur scientists about whether these animals were warm- or cold-blooded. Warm-blooded animals – including humans, all other mammals and birds – tend to have high body temperatures and can tightly control this temperature within a narrow range, generating heat internally as required. Cold-blooded animals – including reptiles such as crocodiles, snakes and lizards, as well as fish and amphibians – have body temperatures that vary with the conditions of the environment they are in, and they tend to have slower metabolisms.

Birds are active, warm-blooded animals with the high metabolism that seems necessary for flight. If this ability evolved somewhere in the lineage running up to birds, it's likely that at least the carnivorous theropods from which they are descended were warm blooded too. Some experts argue that the herbivorous long-necked sauropods weren't warm-blooded, as their massive body size would have caused them to retain all the heat they required. But the pterosaurs, close relatives of the dinosaurs, must surely have had high metabolisms, as they were active, flying animals too, and if *they* were warm-blooded, then maybe all dinosaurs were as well.

'Based on the evidence of posture and locomotion, we actually think that some of these more predatory dinosaurs might have been warm-blooded, whereas some of the larger vegetarian animals might have been cold-blooded', says Michael Novacek a palaeontologist at the AMNH. 'We don't know that directly, but one thing is certain: birds are warm-blooded animals and birds are a kind of dinosaur. So somewhere in the evolution of dinosaurs warm-bloodedness evolved, maybe in those dinosaurs that aren't strictly classified as birds. Maybe *Tyrannosaurus* was warm-blooded. We don't know that for sure, but it's a good supposition.'

A number of different methods have been used over the years to probe fossils for evidence that might answer the question – the microstructure of the bone, for example. In cold-blooded species bone grows in dense rings, but in warm-blooded animals tightly packed cavities called Haversian canals can permeate the structure. Tests of many dinosaur bones appear to show evidence of Haversian canals, perhaps suggesting they were warm-blooded.

Accelerated development

Though the rapid development of young was a method already employed by the non-avian dinosaurs, the ancestors of modern birds used another trick in the timing of their development, which may have been part of the key to the massive success of the group. It may also account for some of the difference in appearance between birds and dinosaurs.

Bhart-Anjan Bhullar at Harvard University in Cambridge, Massachusetts, studies the origins of major vertebrate groups such as birds and mammals, and the unique features that define them. One of his projects involved bird skulls. Bhullar noticed a puzzling similarity between the large eyes, flat faces and rounded shape of bird skulls and those of baby dinosaurs. Rather than having elongated heads with long snouts and jaws as adult dinosaurs do, baby dinosaurs have rounded heads that are relatively large compared to their body size.

'No one had told the big story of the evolution of the bird head before', Bhullar said. 'There had been a number of smaller studies that focused on particular points of the anatomy, but no one had looked at the entire picture ... The origins of the features that make the bird head special lie deep in the history of the evolution of archosaurs, a group of animals [including crocodiles and dinosaurs] that were the dominant, meat-eating animals for millions of years.'

Working with Timothy Rowe at the University of Texas, Bhullar and his team compared adult and baby skulls of birds, dinosaurs and crocodiles. Their results showed that though dinosaur skulls changed in shape dramatically as they matured, adult bird skulls remain remarkably similar to those of juveniles. The conclusion was that the ancestors to modern birds had undergone a dramatic change in their development. In a process known as paedomorphosis or neoteny, they evolved to reach sexual maturity at an earlier stage of development. This meant their adult body shape no longer differed significantly from their baby body shape, and it also meant they concluded their development much more rapidly – in as few as 12 weeks in some species of bird.

Mark Norell of the AMNH was also involved in the work. 'It was actually Mark who brought to our attention the hypothesis of paedomorphosis in birds, pointing out the similarity between the skull of *Archaeopteryx* and the skull of some baby non-avian predatory dinosaurs he'd collected in Mongolia', Bhullar says. 'Our results had already started to converge on this result, and we quickly moved to test it by including some additional relevant specimens. The results were immediately striking.'

Domestic dogs, too, are thought to be paedomorphic versions of wolves, accounting for their low levels of aggression, playfulness, large eyes and puppyish ways. A similar developmental process may also have been at work in the history of our own species. Researchers believe that adult humans retain the juvenile features of apes, such as flat faces and large braincases. In fact, it may have been the benefits of having a relatively larger and more complex brain that made this a good strategy in both humans and birds.

In birds, larger brains may have allowed for more sophisticated visual processing and coordination, both of which are essential for flight.

Bhullar argues that rolling back development to an earlier stage in adults can have some unusual benefits. Because adult paedomorphic animals are typically smaller than their ancestors, it means that, for their size range, they may have unusual traits and adaptations. 'Suddenly a population of small animals will pop up with a set of adaptations totally unlike those of the pre-existing small animals', he says. 'These unique characters may allow the exploitation of radically different ecological niches from other similarly sized organisms.'

It could be that fiddling with their development enabled the massive evolution and radiation of the 10 000 living species of birds – from the diving penguins of the Antarctic to the running ostriches of Africa and minute hovering hummingbirds of the Americas. According to Bhullar, specific features of specialised animals – the kind of things that lock them in to using a particular habitat and prevent them from evolving new kinds of body plan – tend to appear late in embryonic development. This means that paedomorphosis 'has the potential to roll back these features to a more generalised state', unlocking a fresh pool of potential.

Hands to wings

Though the vast majority of evidence from embryonic development points to birds evolving from dinosaurs, until recently something remained that confounded the overall picture of a link between the two. A difference in the arrangement of the bones in dinosaur hands and bird wings had left palaeontologists scratching their heads. That was until *Limusaurus*, a vegetarian theropod with a bird-like beak, was discovered by a team including James Clark, from George Washington University in Washington, DC, and Xu Xing.

The standard ancestral model for a vertebrate hand, foot, paw or hoof has five digits, and most vertebrates still display five digits in the early stages of embryonic development. Any groups of vertebrate that have fewer than five digits have lost some through evolution (horses have taken things to the extreme and now make do with a single elongated toe). During embryonic development, the wings of birds appear to form from the fusion of the three middle fingers of the basic vertebrate hand (that's fingers two, three and four, if you imagine the thumb is one and the little finger is five). Similarly, most theropod dinosaurs, such as *Allosaurus*, have three digits on their hands. This is all well and good – or at least it was until fossils of early theropods, such as *Dilophosaurus*, suggested that the digits of the dinosaur hand were numbers one, two and three. Fossils of *Dilophosaurus* seem to have much-reduced fourth digits and almost absent fifth digits.

Clark and Xu's 2009 study revealed that *Limusaurus* had a reduced first digit as well as other features of its hand similar to those of *Allosaurus*. Taken together, these facts suggested that *Allosaurus* and other later theropods had hands that matched the bone configuration of bird wings after all. These carnivorous relatives of birds 'had digits 2, 3 and 4, but ... these have long been misidentified as digits 1, 2 and 3', Xu told *Nature*.

This problem with digit development was one thing sceptics of the link between dinosaurs and birds had hung onto as evidence that birds were descended from another group. With it, according to Xu, has gone one of the final puzzles clouding the relationship between theropods and modern birds.

The shrinking genome

It's all very well hypothesising that dinosaurs that are closely related to birds were animals with an active lifestyle and a high metabolic rate, but is there any way to test the idea?

The authors of one clever study found a way to infer

metabolic rate in the ancestors of birds. Birds have unusually small genomes (the sum total of their DNA) compared to other vertebrates, a property they share with the mammalian fliers, the bats. In the 1970s, Polish researcher Henryk Szarski hypothesised that small genomes are useful precursors to flight because they allow for a smaller cell nucleus and smaller cells with a larger relative surface area, equalling greater gas exchange and overall increased efficiency.

It sounds almost implausible, as the total mass of DNA in cells is minuscule. Nevertheless, flying is one of the most energy-intensive things an animal can do, so even minor gains can confer an advantage. A 2009 study confirmed the relationship by showing that hummingbirds, which have the highest metabolic rate of all birds, also have the smallest genomes. But how does this relate to dinosaurs? Several years earlier, geneticist Chris Organ, then at Harvard University and now at the University of Utah, came up with an idea for estimating dinosaur genome size.

Organ started to wonder how the small genomes of birds evolved. 'Previous research suggested it was related to flight, because smaller cells are thought to be more efficient', he says. 'But our work suggests the small genomes of birds first appeared long before birds, and therefore powered flight, evolved ... the changes may have started to happen as early as the ancestors of dinosaurs and pterosaurs.'

'My co-authors and I knew that a relationship exists between cell size and genome size, but that relationship hadn't been established in bone tissue', Organ says. 'We thought that if we could establish a relationship between genome size and bone cell size, we could use lacunae – the pockets within bone where cells reside – from fossils to infer genome size.

Organ's team used the relationship between genome size and bone cell size in living species to predict genome size in dinosaurs from the size of lacunae in fossil bones. They applied the technique to the bones of 31 species of dinosaur and found that genome reduction was probably underway 230–250 million years ago, in the ancestors of the saurischian dinosaur group, which includes both theropods and their long-necked sauropod cousins. In contrast, the less bird-like ornithischian group (which includes the duck-billed hadrosaurs and *Triceratops*) did not have such small bone cells and presumably branched off before the genome reduction occurred.

The scientists have found that birds, bats and pterosaurs – the three groups of flying vertebrates – all evolved smaller genomes compared with their close relatives, strongly suggesting it was related to flight.

If we could find dinosaur DNA in fossils we could compare it with the DNA of birds to glean much new information about many aspects of dinosaur physiology. Studies have shown, however, that DNA is very unlikely to last more than 1 million years (see chapter 10), so the chance of finding it in fossils more than 66 million years old is virtually non-existent. Nevertheless, more resilient proteins and even some tissue material have been found in a few fossils, so experts are nowadays paying more attention to the chemistry and other fine details of the fossils they find.

A common misconception about fossils is that they are pure rock, but in many cases fossils retain some original bone material. In 2004, a forensic analysis of bone controversially purported to have found the remains of soft tissue, including blood cells, blood vessels and protein, in a 68-million-year-old *T. rex* bone from Montana. Researchers, including John Asara of Harvard University and Mary Schweitzer of North Carolina State University, subsequently analysed collagen they believed they had found and showed that the sequence of amino acids in it was a close match to that in chicken collagen, providing the first genetic evidence for a link with birds (since the sequence of amino acids in proteins is directly related to the DNA sequence used to produce them). In an effort to silence their critics and prove this wasn't a one-trick wonder, the same team has now extracted soft tissue from the bone of an 80-million-year-old hadrosaur, *Brachylophosaurus canadensis*.

Dinosaur diseases and parasites?

It now seems that dinosaurs were even suffering from some of the same diseases that afflict modern birds. A team including David Varricchio and Jack Horner of Montana State University, as well as Steve Salisbury at the University of Queensland in Brisbane, believes that holes in the jawbones of *Tyrannosaurus* specimens were caused by trichomonosis, a nasty parasitic disease that is endemic in pigeons and fatal to birds of prey. Eagles, hawks and related species can pick up the infection from eating pigeons and pass it on to their chicks. In severe infections, lesions develop throughout the lower jaw and throat and start to eat away at the bone.

Some of the most famous *T. rex* specimens – including the first skull from which the species was described and 'Sue', the world's most complete specimen, housed at Chicago's Field Museum – appear to have characteristic marks of the infection in their jaws. This could have led to swelling, problems eating and eventually death, concluded a 2009 study published in the journal *PLoS One*. 'This finding represents the first evidence for the ancient evolutionary origin of an avian transmissible disease in non-avian theropod dinosaurs', wrote the authors. 'It also provides a valuable insight into the palaeobiology of these now extinct animals.'

It's plausible that some tyrannosaurs may have picked up the infection through cannibalism, but up to 60 per cent of specimens display evidence of face biting in battle, hinting at how the infection may have spread. 'We can see similarities with what has been happening to Tasmanian devils recently, where a debilitating oral cancer is being spread by animals fighting and biting each other's faces', Salisbury says. 'It's ironic to think that an animal as mighty as Sue probably died as a result of a parasitic infection.'

The discovery and sale of 'Sue' is a whole fascinating story in itself. One of the largest, most complete and best preserved *T. rex* specimens ever discovered, she was found by chance in 1990 poking out of a cliff on a ranch near the Cheyenne River Sioux Indian Reservation in South Dakota. More than 90 per cent of the bones are preserved, and even the stapes, tiny bones that connect the eardrum to the inner ear, were found by preparators. At nearly 13 metres long, Sue is one of the largest of the 30 or so *T. rex* specimens discovered so far. The fact that the bones are so well preserved affords the opportunity to examine them for disease and other fine anatomical features, little trace of which are left in most other fossils.

Following FBI impoundment – and a protracted court battle between the commercial fossil hunters who discovered her and the rancher whose land they were on – Sue was eventually sold at auction to Chicago's Field Museum for a figure of US$8.36 million in October 1997. Today this remains the most anyone has ever paid for a fossil, and it is thanks in part to sponsorship from Disney World and McDonald's that Sue ended up in a scientific institution where she could be studied, rather than locked away by a wealthy collector.

Just a week before the auction, at Sotheby's in Manhattan, one commentator showed a remarkable lack of foresight when he told the *New York Times*: 'It's going to cost Sotheby's about a half million dollars to clean those bones, and the dinosaur bone market has been depressed by a lot of recent tyrannosaur discoveries ... They'll never get $1 million, and it's really sad that Sue has come to this.' How wrong he was. Today fossils are regularly auctioned for big money in the United States and there's a thriving black market for specimens illegally shipped overseas from China and Mongolia (see chapter 5), but the auctioning of Sue was where it all began.

Trichomonosis is one disease thought to have been passed down from dinosaurs to birds, but another common affliction that followed the same route is lice. What may be the only genuine fossil louse ever discovered is a 44-million-year-old relative of the feather lice of modern birds, which was dug up from an ancient volcanic crater lake in Germany. Could an ancestor of this species have evolved as a parasite of feathered dinosaurs before the end-Cretaceous extinction event 66 million years ago? A 2011 study led by Vincent Smith of the Natural History Museum in London looked at the DNA of living lice and used it to create a family tree.

Smith's team then used a clever method called the molecular clock. This estimates when species separated from one another based on differences in their DNA and on calculations of the rate at which changes tend to accrue naturally in genes as time goes by. Using what they knew about the 44-million-year-old louse species, they estimated that lice began separating out into different species on different host animals 115–130 million years ago.

But what were they living on? The great evolutionary radiations of feathery birds and hairy mammals that lice infect aren't thought to have come until later. Perhaps these lice were infecting feathery dinosaurs instead. 'Our analysis suggests that both bird and mammal lice began to diversify before the mass extinction of dinosaurs', Smith told reporters. 'And given how widespread lice are on birds, in particular, and also to some extent on mammals, they probably existed on a wide variety of hosts in the past, possibly including dinosaurs.'

Dinosaur diets

In 2012, experts including Phil Currie and Lida Xing announced they had found a specimen of a fuzzy 2.5-metre-long relative of Compsognathus, Sinocalliopteryx, with small birds and another feathered dinosaur species in its gut. They argued that the fact that Sinocalliopteryx was able to catch birds meant it was a highly capable stealth hunter that could strike them before they took flight.

Early birds such as *Confuciusornis* would not have been such proficient fliers as modern birds, however, and probably took longer to get airborne, so they may have been easier to catch.

That *Sinocalliopteryx* specimen had several other interesting characteristics. For a start, the smaller feathered dinosaur it had been eating along with the birds was *Sinosauropteryx*, a leg bone of which was found in its gut (so it had been chowing down on a *Sinosauropteryx* drumstick). It had gastroliths, too, small rounded stones in the gut used by modern birds in place of teeth to help grind down their food. In addition, one of the bones showed evidence of damage from a highly acidic foregut of the same kind used by modern crocodiles and vultures to break down and digest bones.

A specimen of four-winged flyer *Microraptor* found in the prehistoric Jehol ecosystem of north-eastern China had the remains of small Cretaceous birds in its stomach, and experts suggested it may have caught them while climbing and gliding between trees. A *Velociraptor* fossil with flying pterosaur bones in its gut is further adding to our picture of the diet and behaviour of these feathered predators.

More recent work on the crow-sized *Microraptor* show that in addition to the small birds and mammals it snacked upon in the forested, swampy environments of Cretaceous north-eastern China, it also ate fish, as seabirds do today. A 2013 analysis of the largest known fossil of *Microraptor* revealed a wad of partially digested fish bones inside it. The scientists behind that study also pointed to some other features suggestive of a piscivorous lifestyle that hadn't been noticed before, such as forward-angled teeth for skewering slippery fish, and the fact that these teeth only have serrations on one side, perhaps an adaptation to prevent struggling prey from tearing itself apart. Only very few dinosaurs are known to have eaten fish, and these include the spinosaurs, such as *Baryonyx*, which had long crocodile-like snouts adapted for snatching fish out of the water. '*Microraptor* appears to have been an opportunistic and generalist feeder, able to exploit the most common prey in both the arboreal and aquatic microhabitats of the Early Cretaceous Jehol ecosystem', wrote the authors of a study published in the journal *Evolution*.

What we know about diet and feeding behaviour even extends to the feeding style of some species of dinosaur, and here the evidence points to similarities with birds too. In 2013 a team of palaeontologists, medical imaging specialists and mechanical engineers used CT scans and sophisticated computer simulations to reconstruct the muscles, tendons and other soft tissues around the skull and neck of a 9-metre specimen of the predator *Allosaurus*. This enabled them to make the most detailed analysis yet of how *Allosaurus* could move its head and neck, and showed that it had great control, allowing it to tug at a carcass while holding it with its foot, ripping off chunks of flesh in the same style as a falcon or kestrel.

'*Allosaurus* was uniquely equipped to drive its head down into prey, hold it there, and then pull the head straight up and back with the neck and body, tearing flesh from the carcass', said lead author Eric Snively, a palaeontologist at Ohio University. 'Many people think of *Allosaurus* as a smaller and earlier version of *T. rex*, but our engineering analyses show that they were very different predators.' *T. rex* had a much heavier head, and research suggests it dismembered its prey by violently thrashing it from side to side, more in the style of a crocodile than a bird.

Attack of the venomous dinosaurs

There was even a tantalising hint in 2010 that another turkey-sized feathered theropod and close relative of *Microraptor* could have been venomous like a snake (or perhaps more accurately a Komodo dragon, which rather than injecting venom with fangs delivers it via saliva with a bite). Although the *Jurassic Park* movie – which depicted frill-necked, chirping *Dilophosaurus* spitting wads of poison into the eyes of its prey – implied we already knew about venomous dinosaurs, it was largely fanciful and not based on convincing fossil evidence. It was exciting news, then, in 2010, when Chinese palaeontologist Enpu Gong and colleagues of his at the University of Kansas suggested that feathered *Sinornithosaurus* may have been venomous. They reported evidence including long teeth with grooves ideal for delivering the poison, and a space in the skull that could have housed a venom gland.

'*Sinornithosaurus* ... has unusually long [upper] teeth that are morphologically similar to those of "rear-fanged" snakes specialised to carry poison', they wrote:

> This type of fang discharges venom along a groove on the outer surface of the tooth. The mechanism for dispensing the venom may be similar to the system used by open-fanged snakes and lizards that discharge it under low pressure provided largely by force of the bite ... We believe *Sinornithosaurus* was a venomous predator that fed on birds by using its long fangs to penetrate through the plumage and into the skin, and the toxins would induce shock and permit the victim to be subdued rapidly.

This was tantalising stuff, but six months later another team argued in *Paläontologische Zeitschrift* that the long teeth described by Gong's group had just slipped out of their sockets in the specimen they'd looked at, and that the grooves for venom delivery were fairly common features of other related theropods that were not thought to be venomous. 'We fail to recognise unambiguous evidence supporting the presence of a venom delivery system in *Sinornithosaurus*', the authors of the rebuttal wrote. As is often the case in the world of science, it looks like it might be back to the drawing board on that one, but I haven't given up hope that we'll one day find evidence of exotic venomous feathered dinosaurs.

5

Fake fossils

*A nebulous trade in forged and illegal fossils is
an ever-growing headache for palaeontologists.*

A hotly anticipated press conference was held in Washington, DC on 15 October 1999 by *National Geographic* magazine. With much fanfare, they announced the discovery of a new feathered fossil from China that was a chimera with a fascinating mix of characters. A team of palaeontologists, enthusiastic amateurs and editorial staff were behind the naming and description of the species, dubbed *Archaeoraptor liaoningensis*. It was to be unveiled in the November issue of the magazine. In an article covering the crop of feathered dinosaurs discovered in the preceding few years, senior assistant editor Christopher Sloan wrote: 'With arms of a primitive bird and the tail of a dinosaur, this creature found in Liaoning Province, China, is a true missing link in the complex chain that connects dinosaurs to birds.'

But all was not as it seemed. What AMNH palaeontologist Mark Norell later described as an 'unfortunate chapter' in modern palaeontology would foreshadow a growing and serious problem of fraudulent fossils being produced on an industrial scale in China. 'To formally name a dinosaur, or any other species, a description must be prepared and published', wrote Norell in his book *Unearthing the Dragon*:

> For important and high-profile specimens like this one,
> such descriptions often appear in top-ranked international
> scientific journals. The two most prestigious are *Science* and
> *Nature*. Peer review, and pre-peer review, had rejected the
> paper's conclusions and evidence and it never appeared in a
> scientific journal.

So why did *National Geographic* choose to go ahead and publish a description of the species when it had been turned down by academic journals? For a non-academic, popular magazine it was heading into uncharted and risky territory.

The team behind the announcement had no idea on that fateful October day, but within just a few months *Archaeoraptor liaoningensis* would be revealed as one of the biggest fossil hoaxes in history, and the chance discovery of another fossil by Xu Xing was the key to uncovering the deception. *Archaeoraptor* was soon dubbed the 'Piltdown bird' and the 'Piltdown chicken' by the press, in reference to another major fossil hoax in which faked remains of putative early hominids were dug up from Piltdown in England in 1912. For *National Geographic* – a bastion of publishing usually beyond reproach – this embarrassment would be one of the greatest blunders in its 125-year history. But more on that later.

Amateur collectors

The problem of faked fossils in China is serious and growing. It is exacerbated by the fact that most of the fossils are pulled from the ground by desperately poor farmers and then sold on to dealers and museums rather than being found by palaeontologists on fossil digs, which is how specimens are discovered in most other parts of the world.

Liaoning, once known as Manchuria, was annexed by the Japanese as part of the puppet state of Manchukuo from 1931 until the conclusion of World War II in 1945. A few hundred kilometres north-east of Beijing, this impoverished region of China borders North Korea to the east and the Yellow Sea to the south. 'At the hub of China's military industrial engine, its cities are polluted, post-apocalyptic landscapes', writes Norell in *Unearthing the Dragon*. 'The countryside is lush green and stifling hot for half of the year and brown, bleak and cold during the other half.'

Peppered with farmland and factories, Liaoning has been a centre for palaeontological activity since the early 1990s, when many early bird fossils were found there. When *Sinosauropteryx* – the first known feathered dinosaur – was discovered there in 1996, it spurred a fossil hunting gold rush the likes of which had never been seen before.

'[Liaoning] is renowned for the fossils that document, in often vivid detail, virtually the entire biota that lived over a period of several million years during the Early Cretaceous, about 125 million years ago', writes Lawrence Witmer, professor of palaeontology and anatomy at Ohio University in Athens, the United States. 'Although exquisite fossils of diverse vertebrates, invertebrates and plants have been recovered, it's the spectacular feathered dinosaurs that have received most attention and caused much controversy.'

Cretaceous-era Liaoning was rich with lakes and marshes, which – combined with plenty of volcanic eruptions – made an ideal environment for preserving large numbers of fossils, often in great detail. But that's not the only reason Liaoning is producing more fossils than any other part of the world today – China can also invest enormous manpower in recovering fossils. 'Some of these localities are unquestionably very rich in fossils but ... the success is clearly linked to the almost unlimited labour available in China', says Luis Chiappe, director of the Dinosaur Institute at the Natural History Museum of Los Angeles County. He describes the work being done there as the 'palaeontological parallel of the construction of the Great Wall of China'.

Thousands of farmers have become 'bone diggers', who find fossils and sell them to dealers. Although it is illegal, numerous farmers today are involved in digging, which continually yields new species. High-quality fossils can sell for tens of thousands of dollars, so finding one is akin to hitting the jackpot when your monthly earnings total a few dollars or less.

'Most of the fossils in museums in China have been purchased from the farmers or the local people who dig them up', Chiappe explains. 'Some Chinese museums have their own expeditions and go out to collect, but the amount of fossils you can collect in two to three weeks is the same as here [the United States]. I mean you'll be happy if you come back with something, but the bulk of what is collected in China has entirely been dug up by the farmers.'

Xu agrees that many of the specimens from Liaoning have come from farmers and dealers, but adds that fossils he has described from elsewhere, such as Inner Mongolia and Shandong Province, have been excavated by his own team. He doesn't like to buy fossils and has bought fewer in recent years, but he's often faced with a difficult decision: if he chooses not to purchase an important fossil it could be lost forever into a private collection, but if he does purchase it, it encourages farmers to keep on digging.

Having thousands of farmers looking out for fossils is a double-edged sword. Though many more fossils are being discovered, they are collected and prepared in a way that destroys much of the useful scientific information. If scientists don't know which location and rock layers the fossils come from, they can't precisely pinpoint their age and also struggle to confirm their veracity.

Knowing which layers of rock fossils were found in is the key to dating them, based on the geological study of the layering of rocks known as stratigraphy. Older fossils are found in deeper layers, with younger and younger fossils found in successive layers on top. Fossils are typically dated geochemically, by looking at tiny but predictable changes in the radioactivity of elements in the rocks that surround them. Over time, some naturally unstable elements decay into others, and the period of time it takes for half a sample of, say, potassium 40 to decay into argon 40 is known as its half-life (in this case 1.26 billion years). By measuring the ratio of potassium 40 to argon 40, scientists can tell how old a rock layer is. Other radioactive materials and elements can be dated in similar ways, and are also used for radiometric dating. When the same rock layer appears across a whole region, we know that fossils found in it all date to roughly the same age range, which would typically encompass several million years.

Not having good stratigraphic control for the specimens dug up by farmers and sold on to dealers is a major problem. Chiappe gives as an example a study he is conducting on fossils of the early bird *Confuciusornis*, which is one of the most abundant fossils found in Liaoning. His team has measured 180 specimens, and they compare them as though all 180 lived at the same time.

'We treat them as a modern population, but they aren't a modern population', he says. 'They have lived thousands, hundreds of thousands, maybe sometimes even millions of years apart.' If the scientists had data on the precise age of the fossils, they might be able to look into whether the species had undergone changes over time, and with better data on geographic location, they could look at changes between regions. '[But] we don't know that, because we don't know exactly where the fossils come from', Chiappe says.

Sophisticated fakery

Fossils collected without data on the rock layer or precise location are clearly less valuable than fossils collected without this information being recorded. Another much more serious problem, however, is posed by forged, faked and manipulated specimens (such as *National Geographic*'s *Archaeoraptor*) which are becoming increasingly common. Farmers who dig for fossils do so to supplement their meagre, largely subsistence-based living, and they are well aware that more complete and spectacular specimens are worth far more than the fragmentary remains they typically dig up. Some don't even realise they are faking specimens and often combine pieces of different fossils found at the same locale. In the most extreme cases, this manipulation is intentional, and fossils found at disparate locations are joined into single specimens. It sounds crude, but even the experts have to look carefully to detect the trickery when master forgers have been at work.

Fossils can be faked in a variety of ways. Sometimes they're hewn from parts from the same species but come from different individuals, so you might have a *Microraptor* specimen with a skull from one individual, a tail from another and a body from a third. Another method involves combining the parts of different species to make a complete fossil that appears to be a new animal. 'Dinosaurs are very similar to birds, so sometimes these fossils combine different birds, different dromaeosaur specimens, or even birds with dinosaurs', Xu says. But the most extreme kind of forgery takes fragmentary fossils and carves out the missing parts from the stone. '[These forgers] are like artists; they carve the bone from the rock.'

Phil Currie agrees. 'The Chinese are excellent craftsmen and they have a long, long history of this. If part of the specimen is missing, many of these poachers and amateurs in fact will just restore them or mix specimens together', he says. 'If it gets in the wrong place and gets published, then it's a big problem.' Trained experts can detect these forgeries, but it requires hard work and careful examination.

In rare cases fossils are completely manufactured from scratch. Currie saw one example in China while on a research trip with Ji Qiang. 'He got a call that a very nice specimen had been found and it looked like *Archaeopteryx*', he says. 'And so we flew to another part of China to go and look at the specimen based on the photographs, and when he got there, it took just seconds to realise that it wasn't a real fossil at all. It was basically ground-up bone, glued back together in a certain way to look like *Archaeopteryx*, but it was almost 100 per cent manufactured.'

It's a significant hurdle to good science, and one that can't easily be solved. 'Fossil forgery in the last decade highlights some troubling trends in Chinese vertebrate palaeontology', wrote Xiaoming Wang, a palaeontologist at the Natural History Museum of Los Angeles County in an opinion piece for the journal *PNAS*:

> While fossil forgeries unfailingly stoke public fascination
> … the widespread damages that forgery causes are often
> not sufficiently recognised. Amid the renaissance of
> Chinese palaeontology evidenced by stunning discoveries
> of inconceivable riches of fossils, palaeontologic
> science is treading a path never experienced elsewhere:
> commercialisation of fossils and all that goes with a quasi-free
> market of fossil trade that has simultaneously become the
> boom and bane of Chinese vertebrate palaeontology.

The museum boom

As palaeontology has boomed in China so has the museum sector, and new institutions cropping up across the nation have fuelled the market for specimens to fill them. Sometimes these institutions, especially small regional museums, have no trained scientists on staff, and often they're filled with large numbers of fakes alongside real fossils.

In Shandong Province, 100 kilometres south of Beijing, mineral magnate Zheng Xiaoting has used wealth amassed from gold mining to build the largest collection of complete dinosaur fossils anywhere in the world. The Shandong Tianyu Museum of Nature has more than 2300 specimens of early birds (including around

600 examples of *Confuciusornis*) and more than 1000 dinosaur fossils, including hundreds of feathered specimens. The Tianyu collection includes important specimens described in top journals *Science* and *Nature*, such as *Tianyulong*, an ornithischian or 'bird-hipped' dinosaur that appears to have had protofeathers but is distantly related to the theropods, hinting that feathers may have been widespread across the entire dinosaur group. According to Chiappe, however, even fantastic museums such as this are not immune to the problem of fossil fakery. He believes that most of the fossils at Tianyu have been purchased from diggers without proper documentation, so there is no detailed stratigraphic control for any of the specimens and there is therefore uncertainty about their origins.

Based on his recent trips to China, Chiappe believes around 50 per cent of the specimens he's seen in regional museums have been enhanced in some way. 'Sometimes that's not important. It's just a little thing that you can highlight and say, "Well, the left hand was sculpted ... I'm going to exclude this from my study"', he says. 'But sometimes it's more significant, and this is after looking very carefully at the specimens. So if you don't really look carefully at them you could be measuring or doing all sorts of things with them that aren't reliable.'

Anyone working with Chinese specimens needs to have their eyes open to the risks. In the past some scientists have analysed Chinese fossils based on photographs alone, partly because it can be difficult to gain access to the collections, but this is no longer good enough. The differences between real and faked specimens can be subtle, and must be looked for carefully. 'You can't just do this from a photograph that someone sent you', Chiappe says. 'There are so many specimens that have been tinkered [with] somehow that you need to have them under a microscope to verify that they are what they look like.'

An investigative report published in *Science* in 2010 revealed that as many as 80 per cent of marine reptile fossils on display in

Chinese museums had been altered or manipulated in some way. Unfortunately, there are few solutions to the problem of faked fossils in China. Laws that forbid the sale of fossils have stemmed some of the trade (they have harsh penalties – ranging from significant fines to execution – but are rarely enforced), yet much of it continues on the black market.

There are, of course, many truly incredible dinosaur museums in China – undoubtedly among the best in the world, with collections of great scientific value. The museums in Lufeng in Yunnan Province and Zigong in Sichuan Province are among very few international sites that have large open digs with partially exposed fossils, in addition to many reconstructed complete specimens. (Others include Dinosaur Provincial Park in Alberta, Canada, and Dinosaur National Monument in Utah, the United States.)

The museum at Lufeng in particular is truly awe-inspiring. The main display space is filled with more large dinosaur skeletons than you are ever likely to see amassed in one place. These are not models or casts, but carefully reconstructed, largely complete fossil skeletons of about 70 giant reptiles. They're arranged in great herds and atmospherically lit from below, so some seem to float overhead as you wander among the columns beneath. Most of these skeletons are of early prosauropods such as *Lufengosaurus*, ancestors of the giant sauropods (such as *Diplodocus* and *Diamantinasaurus*) that would later reach lengths of more than 35 metres. The fossils displayed at Lufeng's World Dinosaur Valley, about 60 kilometres from Yunnan's capital Kunming, are the dinosaur equivalent of the Terracotta Army. To find even single genuine fossils on display in a museum rather than plaster casts of them is exciting, but when those fossils are largely complete and there are more than 70 of them, it's nothing short of astounding.

The first written references to 'dragon bones' (very probably dinosaur fossils) in Chinese literature date to the Western Jin Dynasty (AD265–371), and refer to their location as the banks of a river in Sichuan. The Zigong Dinosaur Museum, 170 kilometres from Sichuan's capital Chengdu, has more than 30 specimens from a cache of nearly 200 fossils (and 26 species of Jurassic dinosaur) dug up at Dashanpu Quarry.

The Lufeng, Zigong and Tianyu museums are well worth making the effort to visit if you are ever in China.

Fossils flowing overseas

Another element to the illegal Chinese fossil trade is the flow of important specimens overseas. In November 2010 the *China Daily* newspaper reported that, in the preceding three years, China had reclaimed more than 5000 fossil specimens from foreign countries, including Australia, the United States, Canada and Italy. A new law, which came into effect at the start of 2011, levied large fines against any person or organisation moving important fossils overseas without express permission from the authorities. Although there are a few exceptions, most major museums in Europe and the United States have strict rules about acquiring looted fossils. Specimens from China and Mongolia (from where it is also illegal to export them) nevertheless routinely turn up for sale overseas.

A number of high-profile cases of illegal fossil trading over the last few years have brought the issue to the attention of the media. There was a blaze of controversy in May 2012 when a largely complete skeleton of a Mongolian *T. rex* relative, *Tarbosaurus bataar*, appeared for sale at Heritage Auctions in New York. Before the auction, Mark Norell wrote an open letter arguing that the fossil was clearly from Mongolia's Gobi Desert and must have been obtained illegally. Despite an injunction brought by US lawyers in the employ of the president of Mongolia – and a restraining order from a district court judge having been delivered to the auction house that day – the fossil was sold for just over US$1 million.

Norell is certainly in a position to know the fossil was from Mongolia. He's been digging up fossils there for two decades, working alongside the Mongolian Academy of Sciences, and has authored more than 75 papers on his findings. 'In my years in the desert I have witnessed ever-increasing illegal looting of dinosaur sites, including some of my own excavations', he wrote in the letter:

> These extremely important fossils are now appearing on the international market ... There is no legal mechanism (nor has there been for over 50 years) to remove vertebrate fossil material from Mongolia. These specimens are the patrimony of the Mongolian people and should be in a museum in Mongolia.

Although the auction had gone ahead, the fossil was taken into custody on 22 June by US authorities, who seized it from a storage facility. 'We are one step closer to bringing this rare ... skeleton back home to the people of Mongolia', said the president of Mongolia, Elbegdorj Tsakhia, in a statement released by his US lawyer. 'Today we send a message to looters all over the world: We will not turn a blind eye to the marketplace of looted fossils.'

In a strange twist of fate, the leg bone of another *Tarbosaurus* specimen appeared in the window of London auction house Christie's at around the same time. Christie's is just a short stroll from the Natural History Museum, and so Paul Barrett spotted the fossil one day as he was walking past. Immediately suspicious as to its provenance, he wrote to Christie's expressing his concerns. The auction house communicated this to the buyer and pulled the bone from sale.

Norell isn't the only one to have noticed an insidious and growing problem. Phil Currie says he first realised that looting of fossils from protected sites was a serious issue in 2000, when specimens vanished from digs he had worked on for many years in Mongolia. 'I was horrified to see that more than half a dozen skeletons of *Tarbosaurus bataar*, protected in situ by legislation for more than 30 years in the Gobi desert, had been ripped out of the ground', he wrote in a column in *New Scientist* one month after the New York auction in 2012:

> This was the beginning of a trend that each year saw more and more such sites desecrated ... Poaching is as bad a problem to palaeontology as grave robbing is to archaeology. In addition to removing important specimens from their country of origin, the geographic and stratigraphic context of the fossils are lost, seriously compromising their scientific value.

Currie says Mongolian poachers go on the hunt for dinosaur remains that are exposed at the surface of rock faces, then often take a pick-axe to them and work their way through the skeleton until they find the claws and the teeth. In the process, they often destroy the rest of the specimen. Even if these fossils do end up in the hands of museums and academics rather than in private collections, they are next to useless, as the context in which they were found is invariably lost.

Though the *Tarbosaurus* skeleton that had been auctioned in New York was seized by US authorities in June 2012, the legal issues weren't wrangled out until May 2013, when the fossil was returned to Mongolian officials. In a ceremony in a hotel across the street from the United Nations complex in New York, the dinosaur was symbolically handed back to Bolortsetseg Minjin, a Mongolian palaeontologist who had been involved in the fight to stop the auction going ahead, and Oyungerel Tsedevdamba, Mongolia's minister of culture, sports and tourism.

The pair used the opportunity to announce that the fossil would be used as the founding exhibit for Mongolia's first dinosaur museum, the Central Museum of Mongolian Dinosaurs, where Bolortsetseg would act as chief palaeontologist.

In a speech, the minister said that before the controversy over the *Tarbosaurus* remains, Mongolians were vaguely aware of their palaeontological heritage, but didn't have any celebrity dinosaurs to rally around – a situation that was set to change with the fame generated by the skeleton. The creation of a national dinosaur museum and public interest and pride in its exhibits, as well as inspiration for a future generation of Mongolian palaeontologists, is at least a silver lining to the cloud of the looting controversy.

Meanwhile, Eric Prokopi, the Florida fossil dealer who had prepared the *Tarbosaurus* for sale and auctioned it in New York, has pleaded guilty to three charges of illegally importing fossils, and now faces up to 17 years in prison. He was in possession of a number of other illegally trafficked specimens, including duck-billed hadrosaurs, oviraptorids and more *Tarbosaurus* remains. All of these are to be returned to Mongolia, as are the remains that were on sale at Christie's in London, which were sourced back to a British fossil dealer. To avoid litigation himself, that dealer agreed to ship a long list of other fossils back to the United States for return to Mongolia and display in the new museum.

In October 2013, the plot thickened when news sources reported that American actor Nicolas Cage had acquired another Prokopi *Tarbosaurus* skull in a 2007 Beverly Hills auction, paying US$276 000 to outbid fellow actor Leonardo DiCaprio. This fossil is also now likely to be impounded by the US authorities. In January 2014, US media reported that yet another *Tarbosaurus* skull had been seized from the home of a Wyoming fossil dealer.

The significant media interest in the *Tarbosaurus* cases and the illegal fossil dealing has a least brought the issue to wider public attention and should make it much more difficult to auction this kind of material in the future, but there's still a long way to go.

Unravelling the *Archaeoraptor* hoax

After the *Archaeoraptor* fiasco that had proved so embarrassing for *National Geographic*, investigative journalist Lewis M Simons was brought in by the magazine's then editor, Bill Allen, to investi-gate. Simons reported in the October 2000 edition that it was:

> a tale of misguided secrecy and misplaced confidence, of rampant egos clashing, self-aggrandizement, wishful thinking, naive assumptions, human error, stubbornness, manipulation, backbiting, lying, corruption, and, most of all, abysmal communication. It's a story in which none of the characters looks good.

The American part of the story began with the smuggling of a fossil from China to the United States, where it was presented for sale at a major fossil show in Tucson, Arizona, in February 1999. There it was discovered and purchased by Steven and Sylvia Czerkas, well-known palaeoartists and dinosaur enthusi-asts who run a small dinosaur museum in Blanding, Utah. They raised the $80 000 required for the specimen from M Dale Slade, a backer and patron of the museum. The Czerkases were friends of Phil Currie, so they invited him to study the fossil and prepare a publication on it with them in a scientific journal. After an ini-tial glance, Currie, who worked regularly with *National Geographic*, alerted editor Chris Sloan to the fossil. Sloan decided it was the perfect addition to a story on feathered dinosaurs he was writing.

In the November 1999 story that was later denounced, Sloan described his first look at the fossil: 'I've seen feathered dinosaurs specimens, but what Stephen shows me takes my breath away. Its long arms and small body scream "Bird!" Its long, stiff tail – which under magnification erupts into a series of tiny support rods par-alleling the vertebrae – screams "Dinosaur!".' Unbeknown to him at the time, this was because they were from completely different animals.

The Czerkases had hoped to display the fossil in Blanding and that it might be the making of their small museum, but Currie and Sloan persuaded them that in order for the fossil to be studied and for anything to be published on it, it must be returned to China after they were finished with it. Once this was agreed, Xu Xing became involved and was sent from Beijing to examine the specimen before its return to China's IVPP.

Alarm bells started to ring when Timothy Rowe at the University of Texas started to examine the fossil using high resolution X-ray CT scans. These would allow the research team to examine the 3D structure of the fossil, which was set into a cracked slab of sandy-coloured stone. Rowe, a world expert on CT scanning fossils, argued that the specimen had been made from a number of different fossils, and most particularly that the tail did not belong to the body.

Currie agreed that he had some concerns, but the Czerkases refused to believe there was a serious problem with the fossil and they pushed on for publication. Ultimately, both *Nature* and *Science* refused to print a paper on the find. This left *National Geographic* in the awkward position of officially describing a new species, as their print cycle and media machine were already too far ahead to pull the story.

Xu eventually proved *Archaeoraptor* was a fake after happening upon the counter slab of the tail in an institute in China in early 2000. It was attached to the legs of an undescribed dromaeosaur. This proved that the tail belonged to another specimen entirely and had been arranged in a false position in the *Archaeoraptor* fossil.

Cue an extremely embarrassing retraction by *National Geographic*, which was then forced to launch an enquiry and bring Lewis M Simons on board to carry out a very detailed and open investigation into what had gone wrong. Phil Currie would later describe his involvement in this scandal as the 'greatest mistake of my life'.

Subsequent detailed CT scans by Rowe ultimately revealed that *Archaeoraptor* was glued together from 88 different pieces of a number of different fossils. Significantly, two of those were species unknown to science, making the specimens very important in their own right. The tail was from *Microraptor*, then the smallest dinosaur ever discovered (see chapter 7), while the front half was a primitive bird that was subsequently named *Yanornis* in a 2002 *Nature* paper entitled '*Archaeoraptor*'s better half'.

'Now that we know which pieces really do go back together properly and which do not, we can see that there is a new species of extinct bird present in the forgery and that it definitely deserves to be studied and described', Rowe told the BBC. 'The tail came from a different animal altogether, and it has already been described and named *Microraptor*. We may never know where the legs came from.'

Luis Chiappe says it's puzzling how the description of *Archaeoraptor* ended up in print in *National Geographic*, as 'the red flag for that one should have been raised long before it got to that point'. With hindsight it seemed obvious that the animal was a chimera of bird and dinosaur features, he says, but it was put together with great skill.

Xu says that much has changed since then. At the time a forgery such as this was not only unexpected, but also difficult to predict. It's also likely that, in those early years following the discovery of *Sinosauropteryx*, the first feathered dinosaur, people were caught up in a wave of excitement and were perhaps less careful than they might otherwise have been.

'If you look at the background, this is a very complicated story', he says, adding that at many fossil sites in China it's rare to find completely articulated specimens of dinosaurs and birds. 'In most cases when we do fieldwork in Inner Mongolia, in Xinxiang, or other parts of the world, what you find very often is an incomplete skeleton ... You see lots of bones on the surface and you collect those bones and go back to the lab. You need to figure out whether those bones are from one individual or from two individuals or from several individuals. That's a part of the research process for a palaeontologist.'

A lesson learnt

Experts are much more wary of inconsistencies or anomalies in fossils these days, but 15 years ago the assumption might just have been that the specimen wasn't assembled properly or had some elements that had been attached by mistake. 'They prefer to believe that based on the size, the texture of the bones and things like that, they prefer to believe those elements are from the same individual, just assembled mistakenly', Xu says. The assumption would certainly not have been that the fossil had been assembled with the deliberate intention of deceiving people.

'At the time, in 1999, we were not really prepared to face the problem of composite or faked specimens', Xu adds. 'Today, if you see a specimen like that – especially if it's from Liaoning – you will say, "Oh yes, this is definitely a fake specimen", because you know that this is a really serious problem. But a decade ago, people were not prepared to understand and deal with the situation. Sometimes the fakes are not so easy to recognise, and you would rather believe it is something good but not perfect.'

China's new fossil industry has appeared in the blink of an eye and its palaeontological community is still finding its feet, but if Chinese authorities and museums are going to maintain their credibility, they will have to tackle the problem of faked fossils and the trafficking of fossils overseas. A remarkable and ongoing series of finds has given us a window into a fabulously weird and unexpected world, but the trade in faked, manipulated and illegally obtained fossils has tainted what are otherwise spectacular collections.

6

The evolution of feathers

The rise and rise of one of evolution's niftiest inventions.

If you could step back to the Late Jurassic, 160 million years ago, and conceal yourself in the prehistoric foliage of Mongolia, you'd see something remarkable. Between the tree ferns and cycads, an unusual-looking bird would appear. It tidies up a clearing – removing fronds, sticks and other debris. Then, with a dramatic flourish, the pigeon-sized creature stands on tiptoe, puffs up its strikingly coloured plumage, and starts to dance jerkily from side to side, all the while producing clicks and shrill little calls. Most conspicuous are its four long tail feathers, which flick and waft as it shimmies to an internal beat.

This is actually a courtship ritual, very much like the kind played out by birds of paradise today in New Guinea. But on closer inspection, the performer isn't a bird at all. It doesn't have wings, but lightly feathered forelimbs with sharp little claws; and instead of a beak, it has a full set of pointy teeth. What we are spying on is actually a small dinosaur named *Epidexipteryx*, Greek for 'display feather'. A delicately preserved fossil of *Epidexipteryx hui*, featuring impressions of four 20-centimetre-long, ribbon-like feathers, was unearthed in Inner Mongolia in 2007 and described in *Nature* the following year. It was the first clue that feathers found a use in display long before they ever helped a creature become airborne. The scenario described above is fanciful, but the paper's authors, from the Chinese Academy of Sciences in Beijing, are convinced the feathers were used for seduction. 'Ornamental plumage is used to send signals essential to a wide range of avian behaviour patterns, particularly relating to courtship', they wrote. 'It is highly probable that the [tail feathers] of *Epidexipteryx* similarly had display as their primary function.'

There is now good evidence that many carnivorous theropod dinosaurs, even fearsome and well-known species – such as *Allosaurus* and *Tyrannosaurus* – had feathers, and that they used them for a variety of functions. 'The most startling revelation about *Velociraptor* and its kin is that many are now known to have possessed feathers', writes John Long:

> This fact has made us think again not only about the transition to birds, but also about how they might have used their feathers. Did they use feathers in complex mating rituals? Did they use them to brood their young? Or did feathers act primarily as a stepping-stone in the evolution of flight? We know from fossil evidence that some of these scenarios, and possibly all of them, were true.

Feathers are so intimately entwined in our minds with flight that this idea takes some getting used to. Nevertheless, animals with flight feathers can't have appeared from nowhere, so it makes sense that the earliest feathers had another purpose entirely.

The first feathers

Sinosauropteryx was the first dinosaur to be discovered with feathers, in 1996, but since then a great flock of feathered species has flapped or scurried to the fore. The fossils of about 40 species have so far been found either with feather impressions, or with circumstantial evidence in the form of either quill knobs (pits in the bones where the ligaments of feathers attach) or a pygostyle (the shortened tail structure to which a fan of feathers attaches). Nearly all of these feathered dinosaurs are carnivorous theropods, and the vast majority of the new finds come from the Liaoning region of north-eastern China, although some hail from Mongolia and a smattering of finds have been made in Europe, North America and Madagascar (see 'An A–Z of feathered dinosaurs' for a full list).

In the years following the discovery of *Sinosauropteryx*, a handful of other feathered species were found, none of which appeared to have been able to fly – they didn't have fully formed wings or the wings weren't right kind of shape to provide lift. It started to become clear to palaeontologists that feathers may have evolved for another purpose entirely and were only later co-opted for flight.

'The protofeathery fringe of *Sinosauropteryx*, now known to extend to its flanks as well as along its midline, was obviously not made for flight, and it is questionable whether it could have served any kind of function in insulation', mused Kevin Padian in *Nature* in 1998:

> Camouflage, display and species recognition come to mind as other possibilities. *Caudipteryx* and *Protarchaeopteryx* go it one better, evolving long feathers with a central rachis (shaft). Were these feathers airworthy? Their vanes are symmetrical and very even, suggesting interlocking barbs, although most flying birds have asymmetrical feathers. However, the arms of *Caudipteryx* are only 60 per cent as long as the legs ... Evidently, the arms and feathers were not large enough for flight.

The feathers of many of these animals were simpler in structure – more what the experts endearingly call 'dinofuzz' than anything we'd recognise as feathers today. The spread of these species across the family tree does suggest, however, that having feathers was a common trait in the theropod dinosaurs, even among the many species for which no feathers have been recorded.

Despite the fact that feathers were undoubtedly widespread, it took a long time to find dinosaur fossils that preserved any trace of them, since 90 per cent of fossil sites preserve bones alone and no so-called soft tissues such as skin, muscles, internal organs or feathers. For a long time, the only dinosaur-era fossils with feathers were a handful of specimens of *Archaeopteryx*, but as John Long points out, in recent years 'a whole series of great fossils have been discovered of both feathered dinosaurs from a variety of theropod families, right through to many examples of primitive birds. The overall body of evidence from all different directions – growth rates, physiology, bone structure, feathers – points to these dinosaurs being the close relatives of modern birds'.

It's very likely that the earliest feathers didn't function in flight or locomotion, but instead were used like the downy fuzz of chicks for insulation or otherwise for display. 'To start with, feather structures are not all that complicated – they are a coat of simple filaments', says Paul Barrett. 'These animals are small and quite active, they have elevated metabolic rates compared to reptiles, and this is a way of retaining heat in small animals.' The next group of dinosaurs that play with more feather-like structures tend to have big pennaceous feathers (that is, feathers the typical modern shape, with a central vane running the length and interlocking barbs running off to either side), which are more obviously used for showing off. These might be a tail fan or a bunch of feathers on each arm they could have waved at one another or fanned out across the nest to insulate their young.

Feathers as display

Some unusual research into the tail muscles of a bunch of oviraptorid dinosaurs, which were common in Mongolia and China during the Cretaceous, provides perhaps the best evidence yet that dinosaurs used their feathers for elaborate displays. Oviraptorids are a group of parrot-beaked omnivorous theropods, including the Gobi Desert animals that have been found fanned out protectively over their clutches of eggs. Despite not being in the part of the theropod family tree most directly related to birds, they have such a perplexing number of very bird-like features that some people have even suggested they were birds that had lost the ability to fly rather than bird-like dinosaurs. These dinosaurs have a pygostyle, with the final few vertebrae fused to form a ridged, blade-like structure.

A 2013 study by experts including Phil Currie and Scott Persons at the University of Alberta, and Mark Norell at the AMNH revealed something completely new about the display purpose of this structure. In the fossils of five species of oviraptorid – including *Similicaudipteryx* and *Nomingia* – the team found marks on the bones suggestive of a tight group of large muscles that would have allowed the stumpy tail to be flexed and posed in a number of ways. The researchers had found evidence that these dinosaurs had flexible tails they could flick, waft and shake as desired. They argued that the males of these species likely indulged in tail-shaking mating displays to attract females, much as turkeys and peacocks do today.

The fossil of *Similicaudipteryx* in particular has structures on the vertebrae of the pygostyle showing feather attachment and suggestive of a tail fan. The other oviraptorids they studied don't have direct evidence of tail feathers, but as they were close relatives of *Similicaudipteryx* and evolved later, they are likely to have shared the same features. The vertebrae at the base of the tails were short and numerous, which would have made them highly flexible. These animals also had large muscles with many connection points extending along the tail, which could have been used literally to (in the words of the famous Motown track) 'shake a tail feather' – allowing them to flick the tail vigorously both vertically and horizontally. *Epidexipteryx*, which we talked about at the start of this chapter, isn't in the same group of dinosaurs, but there's no reason why it couldn't have used its four long tail filaments (see image section) for much the same kind of mating display.

Moulting

Another benefit of using feathers for display is that, through moulting, animals can rapidly perform a costume change, switching in and out of brightly coloured mating plumage (in the manner of Australia's splendid fairy wrens, the males of which turn a brilliant blue during the breeding season). Other birds, such as those that live in snowy winter environments, can change their plumage for the purposes of camouflage.

Studies of 125-million-year-old *Similicaudipteryx* fossils have also been instructive in this regard; research by Xu Xing and Zheng Xiaoting, reported in *Nature* in 2010, shows a difference in plumage between adults and young. Feathers on the tail and wing of the adult fossil specimens they studied appear to be standard pennaceous quills, but the feathers on the fossil of a juvenile *Similicaudipteryx* show something never seen before in a living bird: while the feather tip has the typical shaft and barbules, the base of the feathers is reduced to a flat, ribbon-like shaft with no barbules.

'This baby dinosaur has bizarre flight feathers, which are strikingly different from those of adults', Xu told *Nature*. For the first time researchers had shown that young dinosaurs had a different kind of plumage from adults, which means that moulting was used to change their appearance. In modern birds, babies would normally simply switch from a downy covering of fluffy feathers to adult feathers without this extra step in between, so something unusual had been discovered.

Not everybody was convinced, though. Ornithologist Richard Prum, an authority on feathers at Yale University, argued that the fossil may simply have captured the feathers in the process of emerging from their 'feather sheaths', which happens during moulting. The jury is still out on that, but the research did show for the first time that dinosaurs underwent plumage changes through moulting.

Important research of this type, which requires many specimens of the same species, has been made possible by the incredible fossil collection built by Zheng Xiaoting, whose Shandong Tianyu Museum of Natural History houses the largest assortment of complete dinosaur fossils – and by far the largest array of feathered dinosaurs and early birds – anywhere in the world.

Across an ocean and a continent, remarkable Canadian fossils, discovered by the University of Calgary's Darla Zelenitsky and her team, provided the second piece of evidence for moulting and plumage changes in dinosaurs. Detailed in *Science* in 2012, these fossils of the speedy, ostrich-like *Ornithomimus* from the badlands of Alberta were also the first feathered specimens ever found in North America, and proved that ornithomimid ('bird-mimic') dinosaurs sported a comprehensive plumage.

In the *Jurassic Park* movie, a herd of ostrich-like, speedy *Ornithomimus* were shown escaping from a marauding *T. rex*. Zelenitsky's work has shown that these animals should have been depicted with feathers rather than the naked, scaly covering they had in the film. The fossils of an adult and a juvenile *Ornithomimus* revealed fluffy filament-like feathers in the juvenile, but long pennaceous feathers on the forelimbs of the adult that wouldn't have been any good for flight. Instead, the researchers think they were used for mating displays similar to the tail fans of the oviraptorids. 'This is a really exciting discovery as it represents the first feathered dinosaur specimens found in the Western Hemisphere', Zelenitsky told reporters. 'This dinosaur was covered in down-like feathers throughout life, but only older individuals developed larger feathers on the arms, forming wing-like structures ... This pattern differs from that seen in birds, where the wings generally develop very young, soon after hatching.'

So there is good evidence of feathers being used initially for insulation and display, but how did they come to be co-opted for flight? Eventually, the extra surface area of feathers on the tail and forearms used for display would have offered some lift when jumping or gliding. Then evolution would have started to select for the running or flying functions of feathers rather than simply keeping warm or showing off. (For more on the origins of flight, see chapter 7.)

Feathered tyrants

Most of the known feathered dinosaurs are close relatives of birds in a carnivorous group known as the maniraptors ('hand snatchers'), which includes dinosaurs such as the dromaeosaurs and troodontids, therizinosaurs and oviraptorids. Nearly all the rest are within in the coelurosaurs, a wider grouping that also includes ornithomimids, tyrannosaurs and compsognathids (see 'Relationships of the theropod dinosaurs'). Although feathers have only been found in a smattering of species across the whole group, the fact that some of them are early and ancestral members (*Sinosauropteryx*, for example), and that known feathered species occur on nearly every branch of the family tree, is strong circumstantial evidence that all members of the group were feathered. The reason that, for the most part, we've only seen feathers on the species from China is that the level of preservation of fossils in Liaoning is exceptional, retaining feathers, skin and internal organs that rarely leave any trace in fossils from other parts of the world.

Until recently, the general consensus among experts seemed to be that *T. rex* and other very large theropod dinosaurs probably only had feathers as juveniles, if at all. The idea was that large animals have no need for feathers for insulation as their surface area is much smaller relative to their body size than that of small animals, and they lose heat to the environment much more slowly (the largest land animals alive today – elephants, rhinos, hippos – are all hairless precisely because they have trouble keeping cool). But the discovery of two feathered relatives of *Tyrannosaurus* has turned this idea on its head.

The first, *Dilong paradoxus* ('paradoxical emperor dragon'), was discovered by Xu's team in Liaoning in 2004. This lightly built predator was an early relative of *T. rex* that stalked the fauna of Early Cretaceous China around 125 million years ago (*T. rex* itself hails from North America and dates to the Late Cretaceous, around 66 million years ago). As it was a relatively small animal, 2 metres long, the downy covering of dinofuzz preserved in the fossil was not wholly unexpected. *Dilong* is among a number of small tyrannosaurs discovered in China in recent years. Others such as *Guanlong wucaii* are also inferred to have had feathers, even though their fossils don't retain direct evidence of this. *Guanlong* (an illustration of which graces the cover of this book) is the earliest known of all tyrannosaurs, at about 160 million years old.

Much more surprising was Xu's discovery of the bus-sized, 9-metre-long *Yutyrannus huali* ('beautiful feathered tyrant') in 2012. Also from the early Cretaceous of Liaoning but around 5 million years younger than *Dilong*, this shaggy predator was closer in size to *T. rex* itself. It showed that simple downy feathers were probably much more widespread among dinosaurs than anyone had expected. *Yutyrannus* is not only the largest feathered dinosaur discovered to date, it is also the largest feathered animal known to have lived; at an estimated 1.4 tonnes, it tipped the scales at about 5000 times the weight of an average bird, such as a pigeon.

Three specimens of *Yutyrannus* were found, two of them with a largely complete set of bones, and they indicated downy feathers on the legs, arms and neck. (The specimens were also found in the same quarry, hinting that *Yutyrannus huali* may have been a social species that lived in groups, and adding credence to Phil Currie's theory that some tyrannosaurs were pack hunters.) A great illustration of *Yutyrannus*, released along with the *Nature* paper that described it, showed a family group in a snowy-looking environment with puffs of breath misting in front of them (see image section). The fact that feathers were only found in a number of patches on the body may just be an accident of preservation, or they could indicate that that they were used only for display purposes rather than as insulation.

In *The Bird: A natural history of who birds are, where they came from, and how they live*, Colin Tudge argues against the idea that only juvenile *T. rex* were feathered, pointing out that if feathers had a sexual display purpose then adults would be even more likely to be extravagantly feathered than their young. 'It is not possible, and perhaps never will be, to state definitively whether *T. rex* did have feathers – or to confirm or disprove the idea that grown-up *T. rex*'s were turned out like burlesque queens. But I reject that school of biology that feels it is virtuous simply to be dour', he writes. 'In reconstructions, *T. rex* is generally dressed in scaly leather, like a Gucci handbag. But it might have been decked in gaudy feathers and, in due season, crested and plumed for good measure.'

Although *Yutyrannus* was still just a fifth the weight of *T. rex*, its discovery certainly increased the chance that *T. rex* might also have feathers, as Tom Holtz, a University of Maryland palaeontologist, told *National Geographic*. And even with a fluffy covering of down it would have been just as scary, he added: 'Underneath the fluff, it's still the same gigantic crushing teeth and powerful jaws and softball-sized eyes staring at you … [feathers] might make it a little more amusing, but only until the point right before it tears you to shreds.'

Filaments, bristles, scales and fluff

A few of the new Chinese fossils hint that feathers have their origins much deeper in the dinosaur family tree, not close to the species that evolved into birds. There's even the tantalising possibility that feathers originated in the ancestors of animals that gave rise to dinosaurs and their sister group of flying reptiles, the pterosaurs.

Among the dinosaurs that have been discovered in recent decades, there have been some pretty weird creatures, from shaggy pot-bellied beasts with metre-long, scythe-like claws to hump-backed predators with fully feathered 'wings'. One of the strangest came not from China but from Germany – from the very same Bavarian Jurassic limestone that offered up *Archaeopteryx* 150 years ago. *Sciurumimus*, the 'squirrel mimic', a small megalosaur with a great big bushy tail of downy feathers, was found in 2012. Head to tail, the fossilised juvenile was about 70 centimetres long, but adults would have been much larger. The really interesting thing about this feathered species is that it is among a very early group of carnivorous theropods and isn't a coelurosaur, the group that contains the majority of the other feathered species we've discovered.

'All of the feathered predatory dinosaurs known so far represent close relatives of birds', said Oliver Rauhut of the Bavarian State Collection for Palaeontology and Geology, one of the scientists behind the find. '*Sciurumimus* is much more basal within the dinosaur family tree and thus indicates that all predatory dinosaurs had feathers.'

There have been other remarkable finds – of seemingly feathered dinosaurs that are even more distantly related to birds. *Tianyulong confuciusi* was a small bipedal herbivore with a fuzzy covering of primitive feathers. Nothing unusual in that, except this animal was in the ornithischian group of dinosaurs thought to have diverged from the line that led to birds 220 million years ago (the ornithischians are one of the two great branches of the dinosaur family, the other being the saurischians, which included the theropods that gave rise to birds and the giant sauropods).

Tianyulong is not the only ornithischian to have been found with structures bearing a similarity to feathers. Psittacosaurus is a very early member of the parrot-beaked ceratopsian lineage that eventually led to Triceratops near the end of the dinosaur era in the late Cretaceous. Psittacosaurus was originally described from Mongolian rock deposits, but more recently discovered Chinese specimens reveal plumes of bristles along the tail. It's therefore possible that all groups of dinosaurs could have had simple feathers. 'There appears to be evidence suggesting that even the filaments of pterosaurs are likely to be a kind of primitive feather', says Xu.

Paul Sereno, veteran dinosaur hunter at Chicago's Field Museum, isn't so sure. He says there's evidence of scaly skin impressions without feathers for large dinosaurs, such as sauropods, and for many groups other than the theropods. There's no evidence of feathers in other ornithischians, agrees Paul Barrett who published a study on this in early 2014 along with David Evans of the Royal Ontario Museum in Toronto. 'We have lots of skin impressions from duck-billed and horned dinosaurs, and none of them show anything that looks like feathers. So it could be that these dinosaurs started off with feathers and lost them, or it may just be some underlying genetic mechanism whereby dinosaurs do stuff with their skin – because they also have lots of armour and spikes that form in the skin too.'

Experts have known for some years now that many pterosaurs had a fur-like covering. 'That's been observed in fossils – there are quite a number of exceptionally preserved fossil pterosaurs, different ages, and under the microscope it seems that they have hair-like structures covering the body', says Mike Benton. 'This was because pterosaurs, like birds, had to have a very high metabolic rate [for flight] and the covering would have been used for insulation.'

The question now is did all dinosaurs and pterosaurs inherit feathers from the same common ancestor, or is it just that the group had a remarkable plasticity that allowed its members to play around with dermal structures of different kinds – such things as bristles, quills, fuzz, fluff, ribbons and, eventually, elegant, complex and beautiful feathers, aerodynamically sculpted for the purpose of flight? 'It could be that all dinosaurs had the propensity to produce feathers or whiskers, but whether we can say that the ones in *Psittacosaurus* are actually feathers, or whether they are some other kind of dermal structures, we don't yet know', says Benton. 'It could be that all small dinosaurs had feathers – it could be lost in the large ones – or it could be that dinosaurs manifested a variety of dermal [skin] growths and dermal structures of many kinds.'

'There might just be a propensity to experiment with skin structures in dinosaurs – and ornithischian dinosaurs just experimented with feathers once or twice', agrees Barrett. 'Maybe they had the underlying genetic machinery but never really went into it for one reason or another. That's one of the big mysteries. We don't know if other groups will have feathers, but it's not looking likely. It's looking like it's part of the theropod dinosaur story, but not part of the story of the other groups.'

Nevertheless, if one of the early ancestors of all dinosaurs – or dinosaurs and pterosaurs – did have simple feathers, then it opens up the possibility of fluffy sauropods as well as fuzzy duck-bills, armoured dinosaurs and their ornithischian kin.

'Some artists have already taken the opportunity and there are stegosaurs and sauropods appearing with hints of fluff', writes British palaeontologist Dave Hone on his blog 'Lost worlds':

At the moment, it's probably best considered not much stronger than informed speculation, but it's certainly not unreasonable as a hypothesis or improbable ... It was thought unlikely to the point of impossibility that ornithischians would have any kind of covering beyond scales and armour, but two different species were found to have filaments less than a decade apart, and it has even been suggested that the famous *Triceratops* had some bristles as part of its skin.

The only skin-impression fossil found for *Triceratops* appears to reveal it had bristles, not purely scaly skin as has been long supposed. The remarkable fossil was found by palaeontologist Bob Bakker and is held at the Houston Museum of Natural Sciences in Texas, where he is a curator (although at the time of writing, nothing had yet been published on it). We know that *Psittacosaurus* had bristles, so given it is an early member of the lineage that led to *Triceratops*, it seems reasonable to assume that these were passed down.

What good is half a wing?

'The sight of a feather in a peacock's tail, whenever I gaze at it, makes me sick!' said Charles Darwin in a letter to his colleague Asa Gray in 1860. The issue troubling Darwin was that a peacock's plumage seemed so exuberantly gaudy and flashy, and so counterintuitive in terms of preventing predation, that it made a mockery of his recently published theory of natural selection. But if he'd spent more time thinking about the structural complexity of feathers, and how they could possibly have evolved through natural selection, that might have led to a sense of nausea instead. Indeed, the evolution of flight and feathers presented such a problem to evolutionary biologists for so long that creationists would routinely cite them as evidence against the process of natural selection. 'What good is half a wing?' has often been asked (and we'll come to the answer in chapter 7).

Darwin's contemporary and friend Alfred Russel Wallace, who came up with the theory of natural selection at roughly the same time and found much evidence for it in his travels through

South-East Asia, was as perplexed as many other evolutionary biologists by the problem. 'Evolution can explain a great deal, but the origin of a feather, and its growth, this is beyond our comprehension, certainly beyond the power of accident to achieve', he told London's *Daily Chronicle* in 1910:

> A feather is the masterpiece of nature. No man in the world could make such a thing, or anything in the very slightest degree resembling it ... Watch a bird sailing high above the earth in a gale of wind, and then remind yourself of the lightness of its feathers. And those feathers are airtight and waterproof, the perfectest venture imaginable!

The idea of feathers and flight are so intimately entwined in our minds that we find it hard to imagine the former might ever have served any other purpose. But it's clear that fully flight-capable animals can't have emerged with a complete set of flight feathers ready to go. Feathers are complex structures and must have developed over a significant period of time.

The feather impressions on the fossils of the 'first bird', *Archaeopteryx*, show that it had flight feathers similar to those found on modern birds, which – something like the wing of an aircraft – are asymmetrical to make them aerodynamic and provide added lift. Despite this, most experts think *Archaeopteryx* must have been a pretty poor flyer that struggled to get airborne and most likely had to make ungainly crash landings. Its lack of a breastbone to which strong flight muscles could attach, in addition to its long bony tail and dense skeleton, denied it the kind of aerial grace modern birds take for granted. In any case, *Archaeopteryx* tells us frustratingly little about the evolution of feathers, even though, at 150 million years of age, it's significantly older than many of the Cretaceous-era feathered dinosaurs discovered in China (a confusing problem that has been labelled the 'temporal paradox' – more on that in chapter 7).

There are several ways of looking back at feather evolution over time – or indeed the evolution of any other trait, including: examining the hard evidence of feathers in fossils; studying the differences between feathers in different groups of animals and using that to make inferences about how they evolved; and looking at the development of feathers in the embryos of modern birds. A slick new field called 'evo-devo' (which is short for evolutionary developmental biology) looks at the changes that happen in the developing embryo and relates them to genetics. The 19th-century German biologist and comparative anatomist Ernst Häckel pointed out that 'ontogeny recapitulates phylogeny'. Ontogeny is the course of development of an organism and phylogeny is evolutionary history. In simple terms, we can, in a sense, replay the history of a feature, such as feathers or teeth, by watching how it develops in the embryo.

In the early 1990s, Richard Prum came up with an idea, based on development in the embryo, about how feathers might have evolved and the steps they went through to arrive at the beautiful and intricate complexity of a modern flight feather. Never did he believe he might one day be able to follow these steps of feather evolution in a series of exquisitely preserved dinosaur fossils, but this is exactly what he has been able to do.

How did feathers evolve?

Prum is the world's pre-eminent expert on feathers. In his early career he travelled the globe from South America to Madagascar observing bird courting and mating behaviours and listening to their many and varied calls. Of particular interest to him were the manakins of the South and Central American tropics, small songbirds with an unusual syrinx or voice box that can produce a diverse repertoire of trills, whistles and buzzes. As a graduate student he began to unpick their complex relationships and produced the first family tree detailing their evolutionary history.

A promising career as a field ornithologist lay ahead of him, or so it seemed, but then in the early 1990s, when Prum was in his mid-20s, calamity struck. Through illness, he began to lose his hearing in first one ear and then the other. Eventually he lost the ability to hear high notes at all – which was disastrous, as it meant he became effectively deaf to most birdsong.

Prum often recounts the story of when he first realised his hearing loss was significant enough to affect his work as an ornithologist. In 1998 he was in Madagascar with a group of University of Kansas colleagues, taking them to see the mating dances of a bird called the velvet asity. The group followed forest trails to find birds Prum had banded four years earlier and quickly spotted one of the males in the same patch he'd seen it the last time he was there. 'I lift up the binoculars, and there he is, and I'm showing all these guys their first velvet asity', Prum told the *Yale Alumni Magazine*. 'He opens up his mouth [to sing] and ... I couldn't hear a damned thing. I knew that my hearing loss had started, but I didn't realise this bird, whose song I had described to science, was now inaudible to me.'

Though disastrous to him personally, this hearing loss was a win to scientific enquiry, as Prum focused the fierce beam of his intellect on another ornithological problem that had never been adequately answered. First he began to look at the colours of birds and the way they use them to attract mates, but he soon began to look more specifically at the mystery of feathers.

Though they are more structurally complex, feathers are similar to hairs, nails, claws and scales in both chemistry and origin. All are made of varieties of the protein keratin and are dermal structures, formed as outgrowths of cells in the outer layer of skin or epidermis.

A typical feather – perhaps much the same kind fabled to have been dropped by Galileo from the Leaning Tower of Pisa to prove that the time taken for objects to fall is independent of their mass, or the kind used as a quill by Shakespeare to scribble down his sonnets and soliloquies – has a central rachis or vane, from either side of which protrude a series of branched filaments or barbs. These barbs themselves also have barbed branches, with tiny hooks that allow them to attach to and align with one another, creating the elegant and neat design that is the modern pennaceous bird feather. Birds also have downy or fluffy plumulaceous feathers, which have no central vane and are a messy jumble of filaments. These are the kind of feathers we use to fill our pillows, quilts and winter jackets, precisely for the reason that they are wonderfully insulating. In fact, a great variety of feathers is built around these models – even down to the eyelashes and whisker-like sensory structures of some species of bird – and all of this diversity comes from a simple hollow tube of protein produced by the skin.

Prum's developmental theory of the origin of feathers looked not to the fossil record and *why* feathers evolved, but instead to the development of the embryos of modern birds and *how* feathers first appear and grow in complexity.

The reason scant progress was made on understanding the origin of feathers before the 1990s is that experts assumed feathers evolved for the purpose of flight, and on that basis struggled to come up with any plausible scenario of how this might have occurred. 'Only highly evolved feather shapes ... could have been used for flight', wrote Prum in a *Scientific American* article he co-wrote with Alan Brush in 2003. 'Proposing that feathers evolved for flight now appears to be like hypothesizing that fingers evolved to play the piano.'

There was also an idea that feathers had evolved from scales, but this seemed all wrong to Prum, as feathers are tubes and scales are flat. Before 1996, scientific discussions about the origin of feathers were entirely speculative, because there were no modern examples of early designs of feathers. '*Archaeopteryx* has feathers that are identical in structure to a modern bird's, so there were no data', Prum says. 'The literature consisted of trying to extrapolate backwards from modern complexity to some simpler, primitive, antecedent structures. Most researchers tried to do that by imagining functional scenarios for the origin of feathers – usually organised around the idea that feathers evolved for flight. These failed completely.'

Instead, Prum proposed a new theory of the evolutionary origin of feathers based on the details of how feathers grow. He developed the theory in ornithology classes he ran at the University of Kansas, but wasn't motivated to publish it until the spate of feathered and fluffy dinosaur discoveries in the late 1990s.

Prum's idea about how feathers evolved is elegant in its simplicity. He argues that the steps via which a single feather develops in a bird today closely follow the evolutionary steps by which the modern feather came about, and therefore offer a window into the past. In the first stage of growth of a feather, the epidermis thickens to form a little bump on the surface of the skin called a placode, which then elongates to create a hollow tube or unbranched quill (stage 1). This then divides into a fluffy tuft of feather filaments with no central vane (stage 2). Following this the feather can develop further branching in one of two ways: it can form a central vane with simple filaments branching off it (stage 3a), or it can form tufts of feathered filaments with no central vane but further branching of each of the filaments or barbs to form barbules (stage 3b). The next step sees both of the former stages combine to create a modern pennaceous feather, and also hooks and grooves form in the barbules, allowing them to grip into and align with one another neatly, a little like a zipper (stage 4). In the final stage of development, the length of the barbs or filaments on one side of the rachis increases to create a modern asymmetrical aerodynamic flight feather (stage 5).

The beauty of this theory is that it's now testable through examination of the fossil record. *Sinosauropteryx*, discovered in 1996, appeared to have had a fluffy covering of simple feathers not unlike the stage 2 filaments; *Caudipteryx*, found a few years later, appears to have this same dinofuzz, but also stage 3 feathers on the forelimbs and tip of the tail with a central vane and symmetrical filaments branching off it. Stages 4 and 5 were found in 2000 in *Microraptor*, a crow-sized theropod with some flight capability, now known to have had blue–black feathers and an iridescent sheen.

Feathers entombed in amber

Although the fossils from China have been a great source with which to test these ideas about feather evolution, even unusually fine shale fossils can preserve only so much detail, and what we can learn from them about feathers is limited. That's why it was so exciting when a team of researchers from the University of Alberta, including Phil Currie, announced in *Science* in 2011 that they had found a veritable treasure trove of tiny feathers in Canadian amber from the Late Cretaceous, 70–85 million years ago.

In a paper illustrated with a beautiful series of colour photos of these feathers and filaments, the authors reported that they had found 11 different types, and that they were likely to be a mixture of bird and dinosaur feathers. You can imagine these long-gone beasts brushing past sticky patches on trees and leaving the feathers behind, ready to be entombed as the sap became amber, preserving the physical structure of the feathers almost perfectly. This was yet another example of something totally unexpected from the fossil record. The researchers had used museum collections to scour 3000 pieces of amber known to have interesting 'inclusions' embedded within them.

The paper's first author, Ryan McKellar, told *The Atlantic* that the discovery was significant for a number of reasons. 'It supports a model for the evolution of feathers that has previously relied on compression fossils that are difficult to interpret and have been hotly debated', he said. 'Also, the amber-entombed feathers show that some of the most primitive feather types, also known as protofeathers, were still around just before the dinosaurs went extinct. They existed alongside feathers that are nearly identical to those of modern birds. Given what we know about the animals that were alive in the area at the time, it is reasonable to suggest that the protofeather-like specimens are attributable to dinosaurs.'

McKellar, Currie and their co-workers reported that the feathers represent a number of distinct stages of feather evolution, from stage 1 protofeathers of the kind sported by *Sinosauropteryx* right through to much more complex types of feather, such as those that help diving and swimming birds repel water and those that birds use for flight. As an added bonus, some of the specimens retained colour, suggesting mottled patterns of brown and black.

In an accompanying commentary published in the same issue of *Science*, Mark Norell noted that it was only a very short while since most people had considered dinosaurs to be scaly and dull-looking animals, and our only peek at feathers from the deep past were at those of *Archaeopteryx*, which appeared so frustratingly modern, offering little clues as to how they had evolved. 'How things have changed – now it would take a warehouse to store all the feathered Mesozoic stem birds and non-avian dinosaurs that have been collected from global deposits', he wrote. 'Feathered animals abound and extend deep into non-avian history – even, perhaps, to the base of dinosaurs themselves. Now, instead of scaly animals portrayed as usually drab creatures, we have solid evidence for a fluffy coloured past.' It may not have been dinosaur blood or DNA, but the discovery of bits of 80-million-year-old feathers preserved in tree resin showed that the idea of dinosaur genes coming from mosquitoes trapped in amber isn't as far fetched as it might first appear.

7

The struggle to the skies

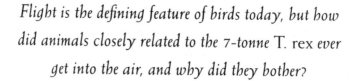

*Flight is the defining feature of birds today, but how
did animals closely related to the 7-tonne T. rex ever
get into the air, and why did they bother?*

One hundred and twenty million years ago, in the Early Cretaceous of north-eastern China, a dinosaur little bigger than a raven
is using his sharp claws to scramble up the trunk of an ancient
conifer. A shimmering golden sliver of morning sun crowns the
distant horizon and bathes the treetops in orange light, illumi-
nating the mist hanging between them. This miniature relative
of *Velociraptor* is covered in long, glossy, blue–black feathers. He
pauses on a mossy branch to preen them, carefully probing his
plumage with tiny teeth and smoothing the feathers down again
with his snout and claws.

Unlike the feathers of some of the larger dinosaurs down on
the forest floor, his are not primarily for display or insulation;
they are fully formed feathers with a central vane and an asymmetrical shape sculpted to aerodynamic perfection. These feathers are intended for flight.

And this little *Microraptor*, as his kind is known, doesn't just
have flight feathers on his forelimbs, but on his hind limbs and

the end of his long tail, too. This early experiment in aviation has four wings and can use them to glide great distances artfully, on a winding path through the forest that takes him from tree to tree.

Preening concluded, a breakfast of fish from the nearby lake plays on his mind. He stands tall and shakes himself, opens his large forewings wide and launches from the branch, his hind wings hanging beneath and behind him. He isn't capable of powered flight by flapping – or at least not very proficiently – but the extra surface area provided by the feathers on his legs and the fan of feathers around his tail makes him a very effective glider.

This *Microraptor* was one of the most common creatures in these Liaoning forests, and more than 300 fossils of his kind have been found so far. Although it was not a direct ancestor of modern birds, this unexpected little dinosaur, and a number of other four-winged flyers, are helping experts understand how dinosaurs first took to the air and made the transition to fully developed birds.

From gliding to powered flight

The first vertebrates to take to the air were probably gliders. Gliding is effectively a form of controlled falling, and is much easier to achieve than powered flight. The difference between gliding and powered flight is the 'flight stroke' of the wings, which generates lift. Rather than having a powerful flapping wing, gliders usually have a rigid gliding membrane, which makes them much less manoeuvrable than true fliers.

True flight not only requires flapping wings, but also strong musculature to power them, and – in birds – a breastbone or sternum to anchor the muscles. Wings (and feathers) are aerodynamically sculpted to provide the greatest lift. Wings tend to be curved on their upper surface and flatter underneath, and are rounded on their leading edge and tapered at the rear. This shape generates upward lift as the wing moves forward through the air, and this lift works against gravity to keep the animal airborne.

Aircraft wings work on exactly the same principle. But whereas aeroplanes use propellers or jet engines to generate thrust that moves them forward, the wings of birds, bats, dinosaurs and pterosaurs have to flap to provide thrust as well as lift. In gliding animals it's falling that provides the thrust that moves them forward and generates some lift that in turn slows their rate of fall; they therefore have to glide at a slight downward angle, and they have to climb to a height to take to the air again, as gliding can only take them downwards.

Birds make flight look easy with their graceful swooping, hovering, turning and diving. The peregrine falcon, for example, can dive at speeds of more than 300 kilometres per hour. The spine-tailed swift can fly in a straight line at 170 kilometres per hour, all the while executing sudden and seamless changes in direction. Hummingbirds can hover on the spot and even fly backwards, while albatrosses and swallows stay airborne for much of their lives, only rarely coming back to land.

Flight may look easy, but in truth it's a marvel of mechanical engineering. Becoming a flyer, and evolving a complex suite of features that will make powered flight possible, is an almost unimaginably complex process. In birds these features include: hands fused into wings, flight feathers, small body size, pneumatisation (hollow bones), low body weight, high metabolism, a keeled sternum and wishbone to which powerful flight muscles attach, large heart, large brain, sharp eyes, powerful visual processing capability – the list goes on. In short, it's so difficult that very few animals have had the evolutionary tenacity to pursue it over the past 350 million years: these were the insects, pterosaurs, birds and bats.

'The transition to flight was a very tough evolutionary step, far greater than our own historical move to walking on two legs', says Paul Sereno. 'There was a great unpopulated niche-land to exploit, waiting for a toothy animal with a backbone that was more nimble than a pterosaur. That's why evolution took them there, but the transition was very, very difficult, and achievable only at small body size.'

Though flight is difficult, the payback comes in some significant advantages over land-bound creatures. Flight allows animals to evade predators rapidly and remain out of their reach. It also allows access to previously inaccessible resources such as food and nesting sites high up in trees, and it allows animals to migrate over large distances in search of food and mates.

Flying may seem energetically expensive in that it uses more fuel than swimming or walking for the same period of time, but it uses less energy to cover the same distance, making it extremely efficient. Walking or running uses up to 10 times as much energy to cover the same distance as flight, and fliers move through the air 10–20 times faster on average than terrestrial animals can move on the ground.

Avoiding predation is a plausible reason that flight evolved in birds, an idea backed up by the fact that if you take predators away, many birds will give up flight altogether. Across the planet, birds have found their way to predator-free islands and, provided they have the resources they need, lost the ability to fly. Most of the examples seem to be in the Pacific – the takahe, kakapo, kiwi and moa of New Zealand come to mind, as well as numerous species of Pacific rail – but there's also the great auk of the North Atlantic, the dodo of Mauritius, the elephant bird of Madagascar and the Galapagos cormorant.

Flight certainly seems to be linked to success, in terms of numbers of species in the groups that have achieved flight. Living birds number around 10 000 species compared to 5500 mammal species. Of those mammals, around 1240 (23 per cent) species are bats. In terms of species, bats are the most numerous of all mammals apart from rodents, although we're rarely aware of it, because bats are active when most of us are asleep. In total about a third of all species of land vertebrate can fly.

'Flight is clearly a factor in the diversity of these animals and how they've been able to exploit the environment', says Luis Chiappe. Insects, which make up the fourth group in which flight evolved, are the most numerous of all animals, numbering perhaps a million species or more. The majority of insects fly during at least one stage of their development, which means that the majority of all animals can fly. There's clearly a strong correlation between flight and evolutionary success. Flight is difficult but very useful. So how did dinosaurs achieve it?

Flight and flightlessness

The sad thing is that once you go flightless, there's probably little chance of evolving flight again. And since flight is a powerful tool to help animals avoid becoming somebody's lunch, when humans spread out across the globe – often with other predatory species, such as rats, cats, dogs and pigs in tow – these flightless wonders, with no innate fear of predators, stood little chance of survival. A study from the University of Canberra has shown that as Polynesians colonised the 269 larger islands of the eastern Pacific around 700–900 years ago, they likely drove around 1000 species of birds extinct, many of which were flightless ground-dwellers.

From tiny to Titanic

What may be the largest dinosaur, *Argentinosaurus*, weighed 100 tonnes and was more than 30 metres long. The smallest member of the dinosaur family is the bee hummingbird, found on Cuba today. Its body is scarcely bigger than an insect's, measuring just 5 centimetres from beak tip to tail tip, and it weighs less than 2 grams. This hummingbird beats its wings 80 times a second, and the energy expended doing this means it has to consume half its body weight in nectar a day, which it has to gather from around 1500 flowers. A single *Argentinosaurus* weighed as much as 50 *million* bee hummingbirds. This staggering figure illustrates the amazing flexibility of the dinosaur body plan, which has adopted more shapes, styles and sizes than just about any other group of animals, but it also suggests that miniaturisation was an essential prerequisite for flight. All animals that have achieved flight started out small, and dinosaurs were no exception. Some of the smallest known dinosaurs, such as *Microraptor* and *Anchiornis*, are those closest to the origin of birds.

'Palaeontologists thought that miniaturisation occurred in the earliest birds, which then facilitated the origin of flight', said Alan Turner of the AMNH. 'Now the evidence shows that this decrease in body size occurred well before the origin of birds and that the dinosaur ancestors of birds were, in a sense, pre-adapted for flight.' Turner was first author of a 2007 *Science* paper showing that the dinosaur lineages leading to birds began to downsize fairly early in their history.

Turner's research revealed an 80-million-year-old dinosaur in Mongolia's Gobi Desert that was an early offshoot of the dromaeosaur lineage (which includes *Velociraptor* and *Microraptor*). *Mahakala* is thought to have weighed 700 grams and measured about 60 centimetres in length. Although the species dates to a period long after the evolution of birds, it is thought to have split off from other members of the group much earlier in the Mesozoic (the era comprising the Triassic, Jurassic and Cretaceous), before the first birds appeared. The experts behind the find said it was evidence that dinosaur lineages had progressively decreased in size as they got nearer to birds. 'Flight isn't an easy thing, because you are, in effect, countering the force of gravity. Being really small appears to be a necessary first step', Turner wrote. 'Other groups that evolved flight, such as pterosaurs and bats, all evolved from small ancestors. With the discovery of *Mahakala* we were able to show that this miniaturisation occurred much earlier.'

Ground-up or trees-down?

There are two competing theories about how the transition to flight was eventually made once dinosaurs were small enough to achieve it: 'ground-up' and 'trees-down'. The issue is still the subject of debate among palaeontologists. The ground-up idea (more widely accepted until recently) suggests that the ancestors of birds flapped their feathered forearms to increase running speed. 'This scenario, to some extent duplicated by some living birds, would explain the origin of flight from the ground up, as these flapping and running dinosaurs became smaller and their wings became larger', Chiappe says.

Some speculate that feathers helped swift predatory species become more streamlined. One theory holds that the sharp retractable claws of the raptors were a specialisation that allowed them to hook onto – and scale the back of – large herbivorous species. In this context, it's easy to see how branched flight feathers could have helped give these animals an extra bit of lift to jump and glide when stalking prey. Another theory suggests that the feathered forearms of dromaeosaurs allowed 'stability flapping' to maintain balance while gripping onto struggling prey, a behaviour that can be observed in hawks and eagles today.

Not everyone agrees, though. David Varricchio at the University of Montana argues that the ground-up hypothesis doesn't really add up. 'Personally, I feel the [theory] is absurd since we have no examples of "running flyers" today that could represent an intermediate stage', he says. 'There are more examples of flying fish than flying runners. But there are many examples of trees-down gliders: gliding frogs, marsupials, various placental mammals, snakes and lizards.' But the problem with the trees-down theory is that although there are hundreds of gliding species alive today, none of them appears to have developed anything like a flight stroke or wing beat that might make them an intermediate stage between gliding and powered flight.

This said, bats may have evolved from gliders and seem most likely to have come from the trees down. Modern bats need a short vertical drop in order to take to the air and are helpless if they find themselves on the ground, needing to crawl awkwardly to a tree or other vertical surface and climb up it to get airborne once more. In outback Western Australia I've seen zoologists throw bats into the air after they've trapped, weighed and measured them, and the bats seamlessly begin to fly once more. In this context it seems likely that bats evolved from gliding ancestors.

Pterosaurs (built on a somewhat similar body plan to bats, with sheets of skin supported by elongate forelimbs) may have evolved from gliders too, but their fossils show no evidence of adaptations to living in treetop environments as bats do. 'If time and again the evolutionary answer to trees-down flight is a membrane, then bird feathers are an exception that suggests a different history', argues biologist Thor Hanson in his book *Feathers: The evolution of a natural miracle.* 'With their unique follicles and helical growth, their complex structure and diversity of forms, feathers seem grossly overqualified for the job. Why go to all that trouble when a flap of skin would do?'

Our old friend Thomas Henry Huxley, Darwin's bulldog, was among the first to posit the ground-up idea, after his studies of *Archaeopteryx* in the 1860s. He suggested it had evolved from a swift bipedal runner, something like *Compsognathus*, which had a very similar-looking skeleton. For the carnivorous theropods, perhaps, the beginnings of flight would have made hunting easier and more effective, allowing them to leap onto prey and run them down, or flap into the air to snatch at insects. Another tick in favour of the ground-up theory is that birds today use their forelimbs to fly, and in order for this to have evolved, it seems they must have been freed up for the purpose in advance. This is exactly the case in a bipedal runner.

In 2003 Liaoning offered up something totally unexpected – and it lent powerful support to the trees-down school of thought.

Microraptor gui was a four-winged species with modern flight feathers on both its forelimbs and its hind limbs, in addition to a fan of feathers on its tail. Xu Xing led a team that described the species from six specimens found in Liaoning, all around 125 million years old. The 80-centimetre-long animal probably held its hind limbs behind it at a 45-degree angle to the plane of its wings, and then glided, like a flying squirrel, between the treetops. Xu's original idea was that *Microraptor* splayed its hind wings out to either side directly behind it, but anatomists have pointed out that this would have dislocated its hip joints. Another idea was that it tucked its hind wings up under its forewings to create a biplane configuration. This has also been discounted, and the best evidence yet, from models flown in wind tunnels, seems to suggest its legs would have hung beneath it but angled out to the sides a little, while its wings (forearms) would have been held out to the side in the position we're familiar with for birds.

Resolving the 'temporal paradox'

One confusing problem in unravelling the evolution of flight has been that most of these feathered dinosaurs are significantly younger than *Archaeopteryx* itself, which is about 150 million years old. This is because not many of the older, Jurassic deposits are as good as the Cretaceous deposits of Liaoning. Although Liaoning's species can give us clues about the transition, they cannot be direct ancestors of birds, as they were contemporaries of a diverse fauna of early birds, which are also found in the deposits. The idea instead is that these species are close parallels of much older animals that sequentially split off on different branches of the family tree on the way to birds. It can be a bit perplexing to think about, but when new types of species evolve, older types often remain alongside them. For example, our own ancestors are known to have included fish and amphibians, but these still exist alongside us today.

Nevertheless, several new dinosaurs with fully developed feathers on their hind limbs and even feet have helped settle this 'temporal paradox'. *Anchiornis huxleyi*, described in *Nature* in September 2009 by Xu Xing and colleagues, is a troodontid species that lived between 151 and 161 million years ago in Jianchang, China. It therefore pre-dates *Archaeopteryx*.

'*Anchiornis* is really amazing. It shows that four-winged dinosaurs, from different groups, lived across different time periods', Xu says. 'We now have quite strong evidence supporting the idea that the first bird was a four-winged animal like *Anchiornis* or *Microraptor*, and it probably originated around 160 million years ago.' Evidence is building that feathered hind limbs were the ancestral condition in a species that pre-dated the split between dromaeosaurs such as *Microraptor*, troodontids such as *Anchiornis*, and birds.

Described in 2011 by researchers including Xu, *Xiaotingia zhengi* is another very bird-like species with feathered hind limbs. It is of a similar age to *Anchiornis* and *Archaeopteryx*, and at the time Xu claimed it knocked *Archaeopteryx* off its perch as the earliest bird. Subsequent studies have disputed this, however, suggesting instead that *Xiaotingia* is an early dromaeosaur.

A study of 11 kinds of early bird — all of which lived around 130 years ago in Liaoning and had long, vaned feathers on their hind limbs — shows that the four-winged flight model continued to be common among early birds after the split from non-avian dinosaurs. This is quite unlike the condition in modern birds that have leg feathers — some varieties of chicken and pigeon, for example. In these species the feathers are small, downy feathers good for insulation but useless for flight.

Using specimens in the vast collections of the Shandong Tianyu Museum, a Chinese team revealed that feathered hind limbs were surprisingly common and may have been an important step that allowed the evolution of full flight. The authors of a paper in *Science*, including Zheng and Xu, noted that these hind-limb feathers were 'aerodynamic in function, providing lift, creating drag and/or enhancing maneuverability, and thus played a role in flight'.

Some of these early bird species – including *Sapeornis*, *Yanornis* and *Confuciusornis* – appeared to be in the process of reducing their hind-wing feathers and developing the kind of scaly, bird-like feet we are familiar with today. This would have allowed them to move more swiftly on the ground, as the hind-limb feathers of four-winged dinosaurs such as *Microraptor* and *Anchiornis* must have been a significant impediment to a speedy escape from predators, particularly if they had to gain height before they could take to the air once more.

Running up that hill

A famous series of studies carried out by Ken Dial, director of the Flight Laboratory at the University of Montana, has offered up some unexpected clues that are instructive in the ground-up versus trees-down debate. Although we may not have found any species that show an intermediate stage between gliding and powered flight, Dial's work has demonstrated what an intermediary stage between running and flying might look like.

Dial had been interested in how flight might have evolved in birds for some time, but found the lack of evidence from the fossil record frustrating. Instead he wondered if he might be able to glean anything from living animals. The species he decided to focus on was the chukar partridge, which is similar in some ways to theropod dinosaurs in that it's a bipedal ground-dwelling runner. It has some flight ability, but chooses to run instead 95 per cent of the time.

Chicks are born in nests on the ground and learn to run rapidly in order to avoid being eaten. To make an efficient escape from predators they have developed something Dial calls wing-assisted incline running or WAIR. Working with his teenage son, Terry, Ken revealed that escaping partridge chicks flap their small wings frantically in such a way as to help create a downwards force as they run up the side of steep inclines or vertical obstacles. This is similar to the way a spoiler pushes a racing car towards the ground to give it traction as it turns a corner. So is it possible that the ground-dwelling feathered dinosaurs first began to evolve the flapping strokes that would later lead to flight as a method to help them rapidly negotiate obstacles in the terrain? It certainly seems plausible.

Since Darwin's day it has seemed idiotic to think that a half-formed wing could be useful. But as evolutionary biologist Richard Dawkins says: 'Half a wing is indeed not as good as a whole wing, but it's certainly better than no wing at all.' What Dial's research suggests is that the stumpy wings of partridge chicks do serve a useful purpose, and that similar proto-wings in dinosaurs may have allowed them to flutter back down to the ground from the trees whose trunks they'd just used their wings to scramble up.

Dial likes to think of his idea as a 'complex marriage' of the ground-up and trees-down ideas. 'We have taken the beautiful sage elements from each one, and I feel we integrated them perfectly to say you never needed to go strictly from the ground up or tree down', he said in 2008:

> The eons-long evolution of flight is revealed to us in the development of baby birds … Our thesis came out from the demonstration of what living animals actually do. And now we have fossils that we never imagined being discovered in China, South America and Africa that look exactly like we expected – dinosaurs with feathers; dinosaurs with half a wing.

Most particularly he points to *Caudipteryx*, a species discovered in Liaoning in 1998. It has a small fan of feathers around its stunted forelimbs, which may have been used for display but could also have been used for wing-assisted incline running.

David Varricchio argues that Dial's research is really interesting, but because partridges evolved from earlier birds that already had full flight capacity, they may not be a fair example to look at when trying to work out how flight evolved in the first place. 'You're basically looking at an animal where its ancestors had full powered flight', he says. 'How do you translate that model back to say this is how a dinosaur evolved in flight?'

There are some other clues that suggest large vaned feathers and wings could have offered benefits for running, leaping and climbing animals long before they became useful for flight. African ostriches have large wings they use as sophisticated air rudders and braking aids to increase their manoeuvrability as they run. This means they can perform rapid zigzag manoeuvres to escape danger, even while running at 70 kilometres per hour. Ostriches are part of an ancient lineage of early birds, and the fact that they are a large flightless species with powerful leg muscles means they can act as a useful analogy for bipedal species of dinosaur. Research on ostrich running mechanics led by Nina Schaller at the Senckenberg Research Institute in Frankfurt, Germany, hints that very large species of feathered dinosaur, such as the 8-metre *Gigantoraptor*, may have used feathered forelimbs to counteract their great bulk as they ran, allowing them to move much more nimbly than we might expect for an animal of their size.

Successful flight

What Ken Dial's research tells us is that we don't have to choose between the two theories: ground-up or trees-down. Perhaps the first fliers developed feathers as an advantage in running and leaping, which also allowed them to run right up into the trees and then glide back out of them to the ground again. Feathers had begun as insulation, gone on to a role in camouflage and display and, finally, much later been co-opted as aerofoils for flight.

We can also see now that the distinction between dinosaurs

and birds is utterly blurred. There is a tight knot of dinosaur species somewhere near the origin of birds: dromaeosaurs and troodontids that had evolved small size, long flight feathers on their arms and legs, and at least some limited flying ability. In some sense what we now define as a dinosaur and as a bird is arbitrary, but many experts draw the line at full powered flight. *Archaeopteryx* is believed to have had this – just – but it still eluded close relatives, such as *Xiaotingia* and *Anchiornis*.

By all accounts, early birds such as *Archaeopteryx* and *Confuciusornis* would have been abysmal fliers, with their heavy skeletons, poor flight strokes and, in the case of *Archaeopteryx*, long bony tail. These traits may have made it more difficult for them to get airborne and rendered them cumbersome operators once in the air. Aerodynamic agility would come later. Nevertheless, these feathered dinosaurs had finally made it to the skies.

This opened up a whole new world of opportunity for them – one that would carry them through the mass extinction that wiped out their land-bound relatives at the end of the Cretaceous, and allow them to prosper and diversify into the most numerous and successful group of vertebrates that has ever lived. As you read this, an estimated 400 billion individual feathered dinosaurs, of 10 000 species, can be found on earth, in almost every habitable environment. You need only step outside and look up into the trees and the wide blue skies to find them.

8

Sex for *T. rex*

From reproductive biology to nesting, brooding and the
act of sex itself, many surprisingly bird-like traits are
turning up in the mighty dinosaurs.

Xu Xing may be a modern-day Indiana Jones, but larger-than-life
Roy Chapman Andrews may have been the inspiration for the
character. In the early 20th century he was an explorer, naturalist
and fossil hunter with the AMNH, leading a series of major expeditions in Asia aimed at finding hominid (early human) fossils.
Images from the 1920s show him astride a horse, clad in khakis
and a broad-brimmed hat. One of his famous catchphrases was:
'Always there has been an adventure just around the corner ... and
the world is still full of corners.'

He was also director of the AMNH from 1935 to 1942 and is
credited with having inspired a generation of kids with his popular books about dinosaurs, prehistoric mammals and the thrilling
perils of adventuring. Many of today's crop of American palaeontologists say their love of palaeontology began with these books.
Not bad going for a man who had to beg for a job at the museum
and started working there as a janitor.

During the 1920s, Andrews led four pioneering expeditions

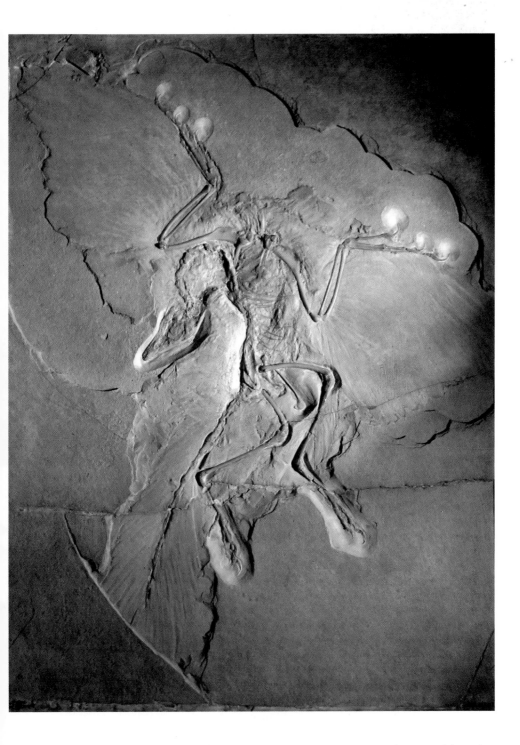

Written in stone. The Berlin specimen of *Archaeopteryx lithographica* is the finest of all 11 known fossils of the species, and the most beautiful. It was discovered in about 1874 and is held at the Humboldt Museum für Naturkunde in Berlin.

SOURCE: H Raab/Wikimedia

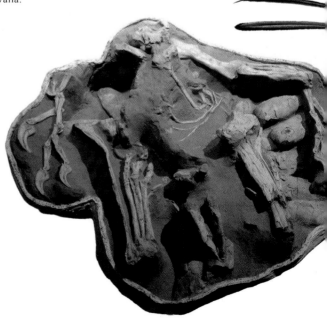

Four-winged flier (*above*). Discovered in 2009, 161-million-year-old *Anchiornis huxleyi* pre-dated the 'first bird' *Archaeopteryx* and helped solve the confusing 'temporal paradox'. Until then, all known feathered dinosaurs were younger than *Archaeopteryx*, so couldn't have been ancestral to it. Instead, experts now think Cretaceous forests were home to a mixture of feathery dinosaurs and early birds.
SOURCE: Julius Csotonyi

Snow stalker (*top right*). A putative Australian tyrannosaur ambushes a group of wide-eyed *Leaellynasaura*. When palaeoartist Lida Xing envisaged this animal for *Australian Geographic* in 2010, he gave it a fluffy covering of polar-style white feathers to reflect the fact it would have been living within the Antarctic Circle on the Cretaceous-era landmass of Gondwana.
SOURCE: Lida Xing/*Australian Geographic*

Baked eggs (*right*). A series of specimens of the oviraptorosaur *Citipati osmolskae* have been found fanned out in bird-like brooding positions atop their nests of eggs in Mongolia's Gobi Desert. Feathery forearms would have allowed these emu-sized dinosaurs to shade their eggs from the heat of the midday sun 80 million years ago.
SOURCE: Dinoguy2/Wikimedia

Fossil flock (*right*). The Early Cretaceous bird *Confuciusornis* was a contemporary of the feathered dinosaurs of Liaoning but was already much more like a modern bird than *Archaeopteryx*, having lost its bony tail and teeth. Thousands of specimens have been found, allowing researchers to study in detail the variation within the species.

SOURCE: Peter Schouten

Dance of death. It was John Ostrom's study in the 1960s of *Deinonychus* that led to the idea they were swift pack hunters with high metabolic rates. Here, in Peter Schouten's illustration, these feathery predators target a *Tenontosaurus*, much as fossils were found locked in combat by Ostrom in Montana.
SOURCE: Peter Schouten

Just for show. *Epidexipteryx*, an odd pigeon-sized dinosaur, was found in China's Inner Mongolia Province and described in 2008. It had a downy covering of fuzz for insulation and four long, ribbon-like tail feathers, which it used for display.
SOURCE: Lida Xing/Zhao Chuang

Handy pose (*above*). Track marks left by the Early Jurassic theropod *Dilophosaurus* suggest it adopted a similar resting pose to that of birds, with the palms of its 'hands' facing upwards – another of many bird traits that evolved in their dinosaur ancestors.

SOURCE: Heather Kyoht Luterman/ Andrew RC Milner et al. (2009)

Night fright (*left*). A new generation of palaeoartists has attempted to capture dinosaurs in a range of behavioural activities and natural situations. Here Alvaro Rosalen shows the suggested threat display of a nocturnal *Epidexipteryx*.

SOURCE: Alvaro Rosalen

On the defensive (*above*). At 8 metres long, the huge oviraptorosaur *Gigantoraptor* is one of the largest known feathered dinosaurs. In this award-winning illustration by Luis Rey, a brooding pair protects its young from an *Alectrosaurus*. Massive nests and eggs (45 centimetres long) found in Late Cretaceous deposits in Mongolia and China likely belonged to *Gigantoraptor*.
SOURCE: Luis Rey

Wonder wings (*top right*). Crow-sized *Microraptor* was the first four-winged dinosaur discovered, in 2000. A number of exquisitely preserved fossils reveal long feathers on the tail as well as on the hind limbs and forelimbs (*see arrows*). There are hundreds of specimens of *Microraptor* in Chinese institutions.
SOURCE: IVPP/David WE Hone et al. (2010)

Food on the fly (*bottom right*). Four-winged flyer *Microraptor* was a contemporary of Cretaceous-era birds and is likely to have preyed upon them, as depicted in this illustration by Australian palaeoartist Brian Choo.

Costume change. Fossils of ostrich-mimic dinosaur *Ornithomimus* revealed in 2012 not only proved the species was feathered, but also suggested that juveniles sported different kinds of plumage from adults. The discovery, led by experts at the University of Calgary in Canada, was the first evidence of feathered dinosaurs in North America and also the first clue that dinosaurs moulted and underwent plumage changes as birds do.
SOURCE: Julius Csotonyi

Snack time. Fossils of *Sinocalliopteryx* with the remains of other species in the gut cavity suggest it was a stealth hunter that fed on the dromaeosaur *Sinornithosaurus* (*above left*) and the early bird *Confuciusornis* (*above right*). SOURCE: Cheung Chungtat/Xing, L. et al. (2012)

Cretaceous clutch (*left*). The nest of an oviraptorid such as *Oviraptor* or *Caudipteryx* from Mongolia. Large capsule-shaped eggs were laid in pairs in a circle, and the dinosaur would have stood with its feet in the centre. It's likely it fanned its feathered forearms across the brood to protect them.
SOURCE: Paleo_bear/Wikimedia

Fight club (*above*). Some dinosaurs may have used their feathered forearms to help give them extra lift – as hinted at here for these duelling *Sinusonasus*, illustrated to magnificent effect by Luis Rey.

SOURCE: Luis Rey

Feathered first (*right*). The little dinosaur that started it all. Found in 1996 and now held at the Inner Mongolia Museum, this *Sinosauropteryx* was the first feathered dinosaur fossil ever found. This specimen (NIGP 127586) is the counter slab, or mirror-image half of the fossil, which farmer Li Yinfang sold to the Nanjing Institute of Geology and Palaeontology.

SOURCE: IVPP

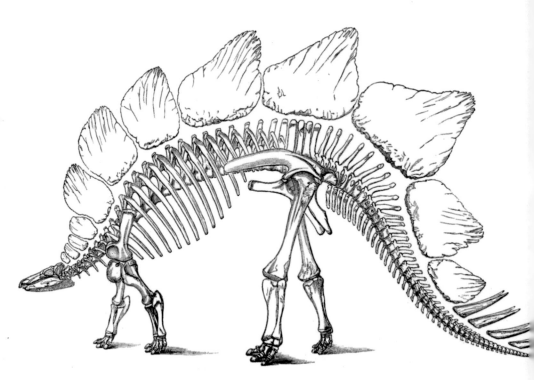

Bones in the badlands (*above*). An 1896 sketch of a *Stegosaurus* discovered at Como Bluff, Wyoming, by Othniel C Marsh. *Stegosaurus* was one of the dinosaurs discovered by Marsh during his 'bone wars' conflict with fellow palaeontologist and archnemesis Edward Drinker Cope.
SOURCE: Othniel C Marsh

Pitstop (*right*). Early birds were contemporaries of a diverse fauna of bird-like dinosaurs, such as *Tyrannosaurus rex*, during the Cretaceous period.
SOURCE: Luis Rey

Border patrol. A 9-metre-long early relative of *T. rex* that stalked the Early Cretaceous of northern China, *Yutyrannus* was the first truly terrifying feathered dinosaur discovered. Its fluffy covering may have helped keep it warm in winter months.
SOURCE: Brian Choo

Family trait. *Tianyulong confuciusi* was an ornithischian dinosaur not closely related to the carnivorous theropods, yet
fossils of it described in 2009 revealed long integumentary
structures likely to have been a form of feather. The discovery of this species suggested that feathers were very widespread across the dinosaur family. Experts now believe that
the flying pterosaurs – a sister group to dinosaurs – may
have had fuzzy coverings of feathers too.
SOURCE: Lida Xing

into the remote Gobi Desert, the first of which was one of the largest scientific expeditions ever to leave the United States. These must have been quite a sight, with great trains of camels and early motor cars slowly snaking their way across the desert landscape. In this windswept and unforgiving region of Asia they found a great cache of dinosaur bones, with the remains of up to 200 individual animals and the first specimens of *Velociraptor* and *Protoceratops*, an early relative of *Triceratops*.

On 13 July 1923 they struck palaeontological gold at a site they'd named Flaming Cliffs, when technician George Olsen found the first evidence that dinosaurs laid eggs. To his excitement, Andrews was shown a fossil nest of 13 eggs arranged in a circle. Each egg was around 9 centimetres wide and twice as long. Although dinosaur eggshell had been discovered in France in 1859, it was misidentified, so Andrews' discovery was the first widely known evidence of dinosaur eggs.

Dispatches sent home to the United States created a wave of excitement about the discovery, and a sketch of Andrews himself appeared on the front cover of *Time* magazine on 29 October 1923. On their return to New York, Andrews was dismayed that all the press wanted to know about were dinosaur eggs, when he felt that other finds – such as the bones of prehistoric mammals – were more significant.

Henry Fairfield Osborn, a student of ED Cope and the president of the AMNH, had employed Andrews to help him gather one of the finest fossil collections in the world, and it was he who described and named many of the species Andrews brought back from Central Asia. Not only had the expedition found dinosaurs associated with eggs, but they had also found largely complete specimens of one emu-sized theropod right on top of the nests themselves. Osborn named the species *Oviraptor* – 'egg thief' or 'egg seizer' – as it appeared to have been caught red-handed plundering the eggs of *Protoceratops*, the bones of which were abundant in the deposits. The idea of a dinosaur brooding its nest and caring for its young was not entertained at the time, although Osborn did leave a little wiggle room when he noted in his description of *Oviraptor* that the name may yet 'entirely mislead us as to its feeding habits and belie its character'.

In 1993, present-day AMNH palaeontologist Mark Norell led another expedition into the Gobi and this time found several more oviraptors fanned out across nests, as well as an embryo fossilised inside an egg. An analysis of the egg showed it to contain a tiny developing *Oviraptor*, not a *Protoceratops*. It confirmed Norell's hunch that these animals were nurturing their own eggs, not stealing those of another species. Pity poor *Oviraptor*, labelled as a thief and devourer of babies, when actually it was a doting parent brooding its nest!

The discovery in the late 1990s of dinosaurs with long feathers on their forelimbs suddenly gave this discovery even more meaning. *Oviraptor* itself has not been found with feathers, but if it had them on its forearms, as seems increasingly likely, then the position of its limbs means it was using them to protect and either shade or incubate its eggs.

In a subsequent paper on the discovery, the scientists argued that the fact *Oviraptor* had been found in the brooding posture taken by living birds, was strong evidence that this kind of bird-like behaviour had first evolved in the dinosaurs. 'It was like finding a snapshot of what the world was like 80 million years ago', Norell told the *Philadelphia Inquirer*. 'Now we know that this nesting behaviour actually predates birds ... It goes back at least 150 million years.'

'This discovery shows us a highly sophisticated behaviour which we thought only existed in modern birds', David Weishampel, a Johns Hopkins University palaeontologist, told the newspaper. 'What this seems to show is probably a mother dinosaur sitting on her nest as a sandstorm comes blowing up and rather than running, she displays altruistic behaviour and tries to protect the nest ... There is no way we could have ever imagined that.'

Devoted dads or hump and run?

This evidence of brooding and nesting behaviour and hints of complex parental care are among a plethora of clues that – in terms of sex, nesting and rearing their young – theropod dinosaurs were very much like birds. The males of up to 90 per cent of modern birds are actively involved in caring for young, which contrasts with reptiles, such as crocodiles, lizards and snakes, where a male's only contribution most of the time is to hump and run – doing little more than fertilise a female's eggs before scarpering.

David Varricchio is one of a number of researchers who have uncovered a wealth of clues about dinosaur brooding and reproduction. He has shown that the bones of some specimens of *Troodon* found fossilised atop their nests contain few of the minerals actively egg-laying female birds might accumulate in their bones. While this doesn't definitively prove they weren't females, combined with the evidence of large clutches of eggs, it suggests the brooders were male, and may hint that the tendency for male birds to care for their young was inherited from the dinosaurs. Today, in the numerous species of bird with large clutches, males typically brood the eggs, which allows females to focus on egg production.

For many years palaeontologists thought we might never know which specimens were female and which were male (not every species demonstrates an obvious size difference between the sexes, and in those that do, it can be either the male or the female that's larger). That was until scientists including Mary Schweitzer at North Carolina State University and Jack Horner at Montana State University cracked open the leg bone of a 68-million-year-old *T. rex* in 2004 and found something inside that looks very much like what's found inside the leg bones of female birds preparing to lay eggs.

Females lay down a tissue known as medullary bone inside

their usually hollow leg bones during ovulation, and it acts as a kind of calcium cache they can draw upon to build eggshells. Once the eggs have been produced, the medullary bone is reabsorbed into the body.

Soon scientists began to find medullary bone in other dinosaurs too, such as *Allosaurus* and *Tenontosaurus* – today its discovery in fossils is not uncommon. Not only had this fortuitous finding given experts a whole new tool for sexing fossil remains, it was also another very strong indicator that birds were the descendants of dinosaurs, as no other animals are known to store calcium in this way.

'Because medullary bone is unique to female birds, its discovery in extinct dinosaurs solidifies the link between dinosaurs and birds, suggests similar reproductive strategies, and provides an objective means of gender differentiation in dinosaurs', wrote Schweitzer, Horner and their colleagues in a 2005 paper reporting the find.

Dinosaur eggs have now been found at hundreds of sites around the world. In terms of the structural make-up of eggshell, they come in three types. The so-called 'spherulitic' eggshells of sauropods and hadrosaurs have spherical patterns in their crystalline structure. 'Prismatic' eggshells have prism patterns in the crystalline structure of the upper layers of the shell. The majority of theropods, including birds, have so-called 'ornithoid' eggs, which once more confirms the close relationship between birds and small meat-eating dinosaurs. In ornithoid eggs, the upper layers of eggshell are made of a spongy mass of crystals, proteins and other organic materials.

Dinosaur eggs come in a whole range of shapes and sizes too, most of which are symmetrical. Many are elongated and almost tube-shaped, while others – sauropod eggs, for example – are nearly spherical. In 2012 some ovoid eggs of a small meat eater were found in Spain that were much more similar in shape to modern bird eggs. Researchers at the Autonomous University of

Barcelona found them in an area of the Pyrenees known to have had coastal deltas, lagoons, beaches and dinosaur nesting sites during the Late Cretaceous, 80 million years ago.

'The size and shape of these eggs with their asymmetric poles are roughly similar to modern hen eggs, which is unusual in the Cretaceous fossil egg record', wrote the researchers in their paper, which went on to say that the dinosaur that produced them probably had a single oviduct like birds. In mammals, such as humans, females have a pair of oviducts attached to their womb. Intriguingly, the scientists suggested that the ovoid shape of modern bird eggs is a consequence of their physiology, and the fact that birds have a single oviduct that produces one egg at a time, rather than the presumed pair of oviducts possessed by their dinosaur ancestors.

In living birds, the isthmus – the part of the oviduct that creates the eggshell membrane – is what produces the asymmetrical shape. Ovoid eggs have a number of benefits: they are less likely to roll out of the nest than spherical ones, and they also allow for an air sac at the wider end from which the bird can breathe towards the end of its development. The fossil eggs from Spain are thought to have been from a relative of *Troodon*, and may be evidence that some dinosaurs shared the same reproductive set-up as birds.

Lizards, crocodiles and turtles lay all their eggs en masse in a single batch, but several things suggest some dinosaurs laid theirs two at a time, one from each oviduct. These include the patterns of eggs in some fossilised dinosaur nests, which reveal pairs of eggs neatly arranged in a circle (see image section), and the 70-million-year-old pelvis of an oviraptorosaur from Jiangxi, China, which has two eggs fossilised inside it.

The greater the number of eggs in the pelvis at any one time, the heavier the burden would have been for flight. The switch from laying many eggs at once to just two is a step along the way to modern birds, which have a single egg in their pelvic canal at a time. Now it seems that some small dinosaur relatives of birds may have laid a single egg at a time too, which along with other adaptations, such as small body size and light bones, gave them a leg-up on their journey towards getting airborne.

Go to your fridge or your chook house and get an egg. Have a good look at it, run your fingers across its surface, and think about the fact that something very similar was first laid by a little chicken-like dinosaur more than 150 million years ago. You could probably have a cracked a few of them into a bowl and whisked them up to make an omelette every bit as delicious as one made from hens' eggs. I wouldn't recommend pinching your breakfast from a *T. rex* nest though ...

Good mother lizards

Arid and remote, Egg Mountain is big sky country south of Choteau in the US state of Montana. The fossils found here are part of the Two Medicine rock formation that has yielded one of the world's largest collections of dinosaur nests, eggs, embryos and babies. It has also been a rich source of adults in association with young, and seems to have been the shared nesting site of several species. Seventy-six million years ago, this active nesting site was situated on a plain lush with ferns, dotted with lakes and bisected by streams. To the west the nascent Rocky Mountains were starting to grow, and 500 kilometres to the east lapped the shores of a shallow ocean.

Today, Egg Mountain is found where the high plains abut the east of the Rockies. 'This gives the site beautiful views of the mountains and shortgrass prairie, but means it is very exposed and we get blasted by serious winds', says David Varricchio, one of the palaeontologists who digs there.

The earliest fossil discoveries here – made in the late 1970s by Jack Horner, of the Museum of the Rockies, and his colleague Bob Makela – were the first complete fossils of baby dinosaurs ever found, along with numerous eggs, embryos and nests. Horner and Makela realised the bones belonged to a duck-billed species of hadrosaur, which they named *Maiasaura peeblesorum*, the 'good mother lizard'. They interpreted the vast number of nests and bones as a great herd of *Maiasaura* killed in a catastrophic volcanic eruption or a sudden flooding event. In addition to that herd-living herbivore were also found the nests of *Troodon formosus* – a small, intelligent meat eater with large eyes. Horner and Makela argued that the site was a great nesting colony that was used year after year for centuries or perhaps millennia.

If you want to picture Egg Mountain at the time of *Maiasaura*, 77 million years ago in the Late Cretaceous, think of great, noisy, smelly colonies of nesting seabirds today – some of these fossil sites would have had hundreds of nests. The fact that some hatchlings appeared to have incompletely developed legs – as some helpless baby birds do today – suggested they were brought food and cared for in the nest and may have remained there, under their parents' watch, for a year or more. At the time the finding was controversial, as parental care went against the idea of dinosaurs as lumbering reptiles, and also aligned dinosaurs much more closely with birds.

The site produced the first dinosaur eggs from North America and they were spectacular discoveries: multiple clutches of the two egg varieties as well as the bones of the ornithopod dinosaur *Orodromeus*. More recent fossil digs led by Varricchio have shown that the eggs belong to *Troodon* and not *Orodromeus*, as had been initially supposed. His work has uncovered many clues about the nesting behaviour of *Troodon*.

From 1984 until 2010 the site sat largely undisturbed without any major excavations, except for a few field schools from the University of Montana, which uncovered important clues about the structure of *Troodon* nests. But then in 2010 Varricchio's team reopened the site in a major way. Rather than jackhammering the vertical face of the rocks they have opened up a large quarry across an area of 70 square metres, which each summer they are working through layer by layer, making important discoveries as they go. They are paying particular attention to coprolites – fossilised scats or dung – and the information that can come from them, such as clues to prey.

Nesting behaviours

A particularly notable trait of modern birds is that many of them nest in trees, or at least above the ground. One theory even holds that birds evolved flight to move their nesting sites into the branches and out of harm's way. There's no evidence of it so far in dinosaurs, but the ancestors of modern birds must have started nesting in trees at some point. Could feathered, tree-dwelling theropods, such as *Microraptor* and *Anchiornis*, have also nested in the trees?

Many primitive modern birds – members of the palaeognath group, including ostriches, emus and cassowaries – nest on the ground today. Some experts believe this hints that this was the ancestral condition for birds, but these birds may just nest on the ground because they're flightless. There's certainly plenty of evidence for ground nesting right across the dinosaur family, but we have no clues as to the nesting behaviour of any of the small feathered dinosaurs or how this might have led to an arboreal (tree-living) lifestyle.

'I certainly think it's possible that dinosaurs could have nested in trees, but it's hard to find evidence for that in the fossil record', Varricchio admits. 'You'd have to have a fossilised tree that fell down ... with an adult and eggs all preserved inside a hollow space in the tree.'

We know that most dinosaurs, including the small relatives of birds, such as troodontids and oviraptorosaurs, buried their eggs in the ground or in nest mounds. '*Troodon* has elongated eggs that are sort of more vertical in the ground, and it seems like only the upper portions of the eggs are exposed. So their eggs are also kind of buried', Varricchio says. This is a more reptilian trait.

One difference between the brooding behaviour of modern birds and reptiles is that birds rotate their eggs to balance out heat transfer. Crocodiles and turtles, however, go through a phase of development where if the nest is disturbed and the eggs are rotated, then the embryos die because they naturally adhere to one side of the egg, and turning them over means development doesn't proceed properly.

'Birds on the other hand have little cords of protein called chalazae that allow the embryo to rotate without affecting it inside the egg, so perhaps the buried condition of dinosaur eggs was representative of a lack of this', Varricchio says. 'Dinosaurs may have been prohibited from nesting in open nests until they had that key feature. So maybe they didn't have what we think of as a typical bird nest up in the tree – eggs just kind of lying in a cup, or a little saucer of sticks and stuff like that. Maybe they actually had to ensure that they didn't rotate by implanting them in the ground.'

Working alongside Frankie Jackson, also of Montana State University, and Darla Zelenitsky of the University of Calgary in Canada, Varricchio has shown that *Troodon* laid its eggs vertically in mud, half-buried them, and then most likely incubated the exposed parts by sitting on top of them and perhaps fanning feathers over them. This is somewhat similar to the nesting style of a peculiar modern bird called the Egyptian plover, which broods its eggs in a nest on the ground where they are half-buried in wet sand. If feathered dinosaurs incubated their nests like birds, it raises another question about their metabolism, because they would have to have been warm-blooded for this to have been a useful strategy.

Increasingly, scientists have been discovering details in the fossil record that just a few years ago we never would have expected to find: preserved soft tissues and proteins, and evidence of colour and feather impressions, for example. Varricchio hopes he might be able to answer the question about tree nesting by keeping an eye out for any circumstantial evidence. To this end, in 2012 he and his team collected data from a modern-day colony of herons, which nest in the trees, to see what kind of debris accumulated underneath and whether any of it could plausibly fossilise. 'We wanted to see if there was any key indication that would reveal that these eggs were blown out of trees, or these eggshells were blown out of trees rather than transported by water or buried in the ground', he says. One day a fossil may be found for which this data proves useful.

Dinosaur milk?

If the thought of a dinosaur perched on a nest in the trees is an odd one, then what about the thought of one feeding its young with milk? Producing milk from a mammary gland – lactation – is the very feature that defines mammals, but it turns out that some birds feed their young in a similar way. These nutritious cottage-cheese-like secretions are produced from the upper parts of the gut and have similar amounts of fat and protein to milk, as well as antibodies, antioxidants and even growth factors. Pigeons, doves, flamingos and emperor penguins all regurgitate 'crop milk' for their hatchlings, suggesting the trait has a long evolutionary pedigree. We already know that many aspects of parental care in birds evolved in their dinosaur ancestors, but Paul Else, a physiologist at the University of Wollongong, has been pondering the idea that crop milk may have helped increase the growth rate of young dinosaurs – something that would have been very useful in species that grew to massive sizes, such as sauropods.

'Dinosaur "lactation" could also have facilitated immune responses, as well as extending parental protection as a result of feeding newly hatched young in nest environments', Else wrote in a paper on his findings. 'Secretory feeding could have been used to bridge the gap between hatching and establishment of the normal diet in some dinosaurs.' He proposes that duck-bills, such as 'good mother lizard' *Maiasaura*, might have initially produced milk and then later regurgitated fermented plant matter to feed their young as they developed.

Enticing displays and terrifyingly large penises

What about the act of sex itself? We've seen from earlier chapters that feathered dinosaurs such as *Oviraptor*, *Caudipteryx* and *Epidexipteryx* used long feathers on their forelimbs and tails to woo potential mates, much as strutting turkeys and peacocks do today. Most mindbogglingly impressive in this regard would surely have been the bus-length parrot-beaked *Gigantoraptor* – one of the largest feathered animals known to have lived.

With a short pygostyle tail dressed with a fan of massive feathers, and possibly with brightly coloured 'wings' too, males of this giant flightless creature surely put on a mating display to rival all others. Perhaps it would have been something like the display of the male ostrich. This bird undulates its body as it dances around, vigorously shimmying with its wings and fluffy white feathers held out to the sides, while also inflating its neck and spiralling its head back and forth. If *Gigantoraptor's* display was anything like this, the earth really did shake when it was ready to bust its moves across the dance floors of Asia 90 million years ago.

Once a male *Gigantoraptor* had drawn in the ladies – and turned away any rivals – with his flashy display, he would have to turn his mind to copulation itself. All large dinosaurs would have had to approach the act of sex with some care.

Some scientists have argued that the very largest dinosaurs couldn't have mounted one another from behind because of the great weights involved and physiological constraints. But Adelaide University's Roger Seymour has studied blood pressure in giraffes – the only living animals with a neck remotely analogous to that of a giant sauropod. To pump blood up their 2-metre-long necks, giraffes need blood pressure seven times the mammalian average, so you can only wonder what kind of pressure *Diplodo-cus* or *Argentinosaurus*, with necks perhaps five times that length, must have required. Seymour's studies of giraffes suggest that, for sauropods, rear mounting wouldn't have been a problem as long as they held their necks horizontally. One intriguing idea is that, once mounting had been achieved, the curiously tiny arms of *Tyrannosaurus rex* might have been used by males to help them hang on to females.

'For very large dinosaurs, weighing tens of tons, as do ele-phants today, it must have been a somewhat delicate business', writes John Long in his book *The Dawn of the Deed: The prehistoric origins of sex*, one that involved 'balancing vast weight on the female's back and certainly involving male reproductive organs that evolved on a truly grand scale, perhaps matching or exceed-ing those of the living whales'.

We can make some inferences about the reproductive equip-ment of dinosaurs from their living relatives – reptiles and birds. Crocodiles and turtles have penises, and most other reptiles sim-ilarly have some kind of organ for copulation, although in many it's a paired structure in which each of the two parts is known as a hemipenis. Tortoises have comparatively enormous penises because their large and awkwardly shaped shells prevent the male from getting very close to the female during sex.

The males of many birds do not have a penis; these birds simply press their multipurpose genital openings together in a 'cloacal kiss', and the male transfers sperm to the female (the evoc-ative name cloaca comes from the Latin for sewer). Speed and effi-ciency are the name of the game in bird sex, rather than pleasure and satisfaction – particularly for a Eurasian songbird called the dunnock, which can mate on the wing in less than one second.

It seems the lack of a penis may not have been the ancestral condition for birds, however. Primitive flightless palaeognaths, such as ostriches and emus, and a related group, including ducks and geese, do have penises. 'Ancient birds probably did possess a penis, so it's quite likely their previous closest ancestors, the meat-eating theropod dinosaurs like *Tyrannosaurus*, also mated using an eversible penis', Long writes. 'Most likely a terrifying large one.'

For a 12-metre *T. rex* to mate effectively, Long believes its penis would need to have been around 2 metres long, and this certainly seems plausible considering the penises of living blue whales are 2–3 metres long (blue whales produce up to 20 litres of semen upon ejaculation). This penis would have been eversible, remaining coiled up inside the genital opening when not erect, as it does in many animals, including ostriches, turtles and crocodiles, today. The corkscrew-shaped penises of ducks are explosively eversible and can emerge to reach their full length of about 20 centimetres in a fraction of a second. I have to agree with John Long that the thought of a 2-metre-long, explosively eversible, corkscrew-shaped *T. rex* penis is pretty terrifying.

Other experts concur that dinosaurs probably had penises, and that they would have been hidden most of the time. 'With nothing hangin' out in the breeze', male dinosaurs, similar to birds and reptiles, would have been difficult to identify, writes US palaeontologist Kenneth Carpenter. 'Assuming you were stupid enough to sneak up under a *T. rex* and pull the cloaca open, the last thing you would ever see during the last moments of your life would be a penis if it was a male, probably similar to that seen in a crocodile.'

These ideas remain little more than titillating extrapolations until more evidence is found. Given, however, that more and more dinosaur fossils from China and Germany have incredible levels of soft tissue preservation, revealing feathers and internal organs such as the intestines and liver, Long believes a fossil penis may one day be found.

In reality, the 10 000 living birds have a great variety of mating practices and strategies – from nesting and brooding to courtship displays and genital structures. We know that dinosaurs had a great variety of body shapes and lived their lives in a multitude of ways too, so it seems likely that this kind of diversity applied to their no doubt colourful sex lives as well.

9

Colouring in the dinosaurs

*A series of clever studies has painted prehistoric worlds
with unexpected colour and it may yet start to fill
them with sound, too.*

Ninety-four million years ago a vast inland sea runs the length
of North America from the Arctic Ocean to the Gulf of Mexico.
Its waters are an effective barrier that divides the landmasses of
Laramidia to the west and Appalachia to the east, and both are
home to a diverse and unique fauna of early birds and feathered
theropods. Predatory reptiles stalk the waterways too, and among
their number is a lone ichthyosaur of a kind known as *Platypterygius*. Shaped something like a fat dolphin, this 7-metre-long air
breather is highly adapted to life in the oceans. She has massive
saucer-shaped eyes that allow her to pick up traces of light in
the depths where she hunts for squid, and she gives birth to live
young in the water, so does not need to venture onto land to lay
eggs as turtles and crocodiles do.

Hunting deep-sea squid is not the only similarity she has with
modern sperm whales. Like them, she is also a uniform jet black,
which helps camouflage her in the inky depths and absorb as
much warmth from the sun as possible when she comes to the

surface for air. Normally she spends as little time near the surface as she can, as this is where she is most vulnerable to larger predators. Today, though, she is lingering, as she is about to give birth.

This *Platypterygius* is among the last of her kind. Ichthyosaurs have been enormously successful for 150 million years, but they have been on the decline for tens of millions of years and now they face a new threat in the form of larger marine reptiles known as mosasaurs. Little does she know it in her current agitated state, but a 15-metre-long mosasaur has been circling in the waters beneath her for some time now. Part of a group that also gave rise to monitor lizards and snakes, mosasaurs have elongate bodies, four flippers and long tails with broad flukes on the end. This predator is darkly pigmented on its upper side and lightly pigmented on its underside, making it difficult to spot from both above and below. Before the *Platypterygius* has the chance to give birth, the mosasaur shoots up from the depths, violently clamping her within its wide jaws, and bringing nearer to its end the long tenure of ichthyosaurs in these prehistoric waterways.

This scenario is fanciful, but thanks to a study published in early 2014 by researchers led by Johan Lindgren at Lund University in Sweden, we do now have some good clues as to the colour of both ichthyosaurs and mosasaurs. The research was the first to reveal the colour of extinct marine creatures and it followed on from a string of papers in recent years that had revealed the colours of prehistoric birds and feathered dinosaurs by mapping pigment-bearing structures within fossilised skin and feathers.

'This is fantastic!' Lindgren told reporters. 'When I started studying at Lund University in 1993 ... it was unthinkable that we would ever find biological remains from animals that have been extinct for many millions of years, but now we are there and I am proud to be a part of it.'

In the 1990s, any book, teacher or scientist would have told you we'd never be able to tell anything about the colour of extinct animals such as dinosaurs. The best we could ever do was make comparisons with living creatures, and often reptiles were deemed to be the best analogies for dinosaurs. Most dinosaurs were painted in similar shades to crocodiles or monitor lizards, in greys, greens and browns. It seems puzzling now when you think about it. Birds and mammals – from flamingos, peacocks, parrots and cassowaries to tigers, zebras, baboons and red pandas – are a large range of vibrant shades and hues, and even many lizards – from chameleons to anoles – have striking patterns of colouration.

Since the discovery of feathered dinosaurs in 1996, the assumption had been that they might have had some of the same vast variation in plumage that birds do today, but few people believed such a thing would ever be confirmed.

That was until January 2010, when a remarkable paper in *Nature* suggested that *Sinosauropteryx* had sported ginger and white stripes around its tail, perhaps something like the pattern found on ring-tailed lemurs today. 'Oh no, it's Ginger-saurus! For first time scientists uncover colour of dinosaur and it was … a red-head' was one headline, in the United Kingdom's *Daily Mail* newspaper, which went on to say: 'As if its short stature and ugly feathers weren't enough to give it an inferiority complex, one of the world's best preserved dinosaurs now turns out to have been ginger.' A jaunty illustration painted by Chuang Zhao and Lida Xing, and released when the discovery was announced, depicts two cheeky-looking, ginger-fluffed *Sinosauropteryx*. Their heads are thrown back, and each is cavorting on a single leg with arms flung wide to impress or perhaps intimidate the other.

The research study, from scientists in China and the United Kingdom, also revealed black, white and orange–brown colouration on the early bird *Confuciusornis*. Similar work has now revealed the feather colours of *Archaeopteryx* and of the four-winged flyers *Anchiornis* and *Microraptor*. These discoveries have opened up a novel field of research, allowing palaeontologists to delve back more than 100 million years and probe the lives of dinosaurs and early birds.

'Feathers are key to the success of birds and we can now dissect their evolutionary history in detail', Mike Benton, one of the experts behind the work, told reporters. 'The simplest feathers, in dinosaurs such as *Sinosauropteryx*, were only present over limited parts of its body – for example, as a crest down the midline of the back and round the tail – and so they would have had only a limited function in thermoregulation [maintaining body temperature].'

His team therefore suggested that feathers arose initially for colourful display purposes and only later were co-opted for insulation and eventually flight. The idea that display and communication were the initial functions of feathers is interesting because most experts believed the first feathers were for insulation. If colourful feathers evolved for display, they might have played a much more integral role in the success, evolution and diversification of dinosaurs than has been supposed.

Uncovering the colour of feathers

How on earth can you find out anything about the colour of feathers from the fossil record? I admit to having been baffled by this when stories of the orange plumage of *Sinosauropteryx* first broke in 2010. To answer that question, we have to look into a little of the science of how animals make colour in the first place.

Pigmentation

The pigment that gives our hair and skin colour is called melanin (from the Greek *melanos*, meaning 'dark') – it's the same substance that is produced when we sunbathe, making us look tanned. Like all pigments, melanin works by absorbing some wavelengths of light and reflecting others to produce specific colours. Inside hair and feathers, it's wrapped in tiny packages known as melanosomes, which create shades of black, grey, orange and brown. When I say tiny, I mean *really* tiny; most are 200–600 nanometres (millionths of a millimetre) across – 200 of them can fit across a human hair.

Though diminutive, they are incredibly tough and actually form part of the strong protein structure of hair and feathers. They are so tough, in fact, that they can survive in fossils for hundreds of millions of years.

According to Mike Benton, until very recently, people just wouldn't have believed melanin would get preserved. Even more so after a series of high-profile failed efforts to recover DNA from dinosaur bones in the 1990s made people extremely cautious about attempting to retrieve proteins or any other organic molecules from fossils. But it turns out that melanin is a very tough chemical, and part of its function in hair and feathers is to make them strong. 'This is why, when you get older, and like me your hair gets grey, it actually gets weaker', Benton explains. In any case, he says, 'we're not detecting the presence of melanin by chemical means in fossils, we're doing it by physical means. It's because it's encapsulated in these melanosomes'. Keratin – the protein from which hair and feathers are made – is a plasticky kind of substance, so in order for the melanin to get into it, it needs to be encapsulated.

Jakob Vinther, a molecular palaeobiologist based in Benton's department at the University of Bristol, was the mastermind behind the colour-identification technique, in which different pigments, such as red, brown, buff, grey and black, are detected simply by looking at the shape of melanosomes in the fossils using a powerful electron microscope. In modern birds, melanosomes that result in different feather colours are different shapes: while sausage-shaped eumelanosomes contain the pigment eumelanin and create black plumage, the spherical phaeomelanosomes contain phaeomelanin, which creates orange plumage. These shapes are what Vinther first searched for in fossils of the early bird *Confuciusornis* while he was still a graduate student at Yale University.

In mammals, pigment is the common way to produce and display a colour. Pigments in both plants and animals work by reflecting and absorbing different wavelengths of light. White reflects all wavelengths of light, black absorbs all wavelengths, and colours in between selectively absorb some wavelengths but not others. Chlorophyll, which makes plants green, for example, absorbs all red light but reflects green; anthocyanins, found in red leaves, absorb green and blue light but reflect red and yellow.

In 2013, a study that simulated the fossilisation process cast some doubts over the reconstructions of dinosaur colour from melanosomes. Maria McNamara, a researcher then in the same department as Vinther and Benton at Bristol University, attempted to mimic fossilisation by subjecting modern feathers to great heats and pressures akin to those they might experience underground in the earth's crust. 'A brief spell in an autoclave can reasonably simulate the effects of temperature and pressure during burial over millions of years', she told *Nature*. Her research indicated that melanosomes shrink during fossilisation, which might affect their shape and therefore the reconstructions of their colour. Vinther's response was that the melanosomes shrank almost equally in several dimensions, so his reconstructions of colour shouldn't be affected. Only time and more research will tell whether the fantastic new visions of coloured dinosaurs are accurate or not.

Iridescence

Birds, insects and fish, however, have another trick besides pigment up their sleeve, and it allows them to be decorated in a much brighter, brasher range of gaudy hues than we comparatively drab mammals could ever hope for. Iridescence is so-called structural colour, which occurs when light bounces off physical features in the surface of feathers or scales and is split into different colours, in much the same way a prism splits white light into its constituent colours. Some birds – parrots with green plumage, for example – use a mixture of both yellow pigments and blue iridescence to create their colour.

The structural features that manipulate the light vary, but

include wafer-thin stacks of translucent organic material that interferes with and reflects light. These films, made of chitin in insects, can reflect and amplify light of one particular colour or wavelength over and above others. This is how iridescent or metallic hues are produced, such as those that adorn the feathers of birds of paradise and peacocks, butterfly wings and a whole spectrum of beetles. Sir Isaac Newton, who shed much light on optics and refraction, was the first to reveal that minute layered structures were the cause of colour in peacock feathers.

In 2003 I first learnt about the possibility of structural colours persisting in fossils when I wrote a story for National Geographic's website about a 50-million-year-old beetle fossil that still had a brilliant-blue iridescent sheen – in this case, the fossil was so incredibly well preserved that the 'multilayer reflector' that created the colour in the surface of its exoskeleton was still intact. At the time the fossil was the oldest known to retain any bright colour, and it may still hold that record. Oxford University's Andrew Parker said then that we might one day be able to study the physical features of dinosaur fossils to predict what colours their feathers might have been. It seemed like science fiction to me, so it was a thrill to see the idea come to fruition just seven years later.

Vinther, Benton and their co-workers at Bristol University have been able to find evidence of these structural iridescent colours in fossilised dinosaur feathers by looking at the density, orientation and stacking of the melanosomes. In some cases, melanosomes act to produce colour in two ways: through the pigments wrapped up inside them, but also through their stacking and organisation, which interferes with and manipulates the light that hits them.

Sinosauropteryx was just the start, and now a series of compelling papers has detailed the colours of Archaeopteryx, Anchiornis and Microraptor. National Geographic described Anchiornis as having 'looked something like a woodpecker the size of a chicken, with black-and-white spangled wings and a rusty red crown'. The team behind that discovery, including Yale University feather expert

Richard Prum, analysed the colour on 29 different regions of the animals' body, giving them a largely complete picture of the overall plumage pattern. *Archaeopteryx* would have been black, while a 2012 study of 130-million-year-old *Microraptor* revealed it would likely have had dark-blue to black plumage with an iridescent sheen – perhaps similar to a crow or raven.

'Modern birds use their feathers for many different things, ranging from flight to thermoregulation to mate-attracting displays', Prum's co-author Matt Shawkey, from the University of Akron in Ohio, told reporters. 'Iridescence is widespread in modern birds and is frequently used in displays. Our evidence that *Microraptor* was largely iridescent thus suggests that feathers were important for display even relatively early in their evolution.'

A third method is being developed that actually looks for remaining chemical traces of pigments themselves, based on the fact that when eumelanin forms it binds up copper and so in fossils leaves telltale traces of the metal behind. 'There has been proposed a method to look at pigments using metal mapping of copper', Vinther says, but it would only be useful for detecting black melanin, and – because copper levels fall as the melanin degrades – it would also be prone to error from other sources of copper in the deposits. It remains to be seen how effective this method will be, but it's exciting in prospect.

The value of colour

The fairy wrens of Australia and New Guinea are delightful little songbirds, some of which have brilliant blue and iridescent plumage. Recent studies have also shown that male fairy wrens of some species have patches of feathers that reflect ultraviolet (UV) light. This is invisible to us, but appears as another layer of colour to female wrens. While humans have three types of colour-detecting rod cells in their eyes – red, blue and green – birds have a fourth, which detects UV light. This means they see the world in a much more complex palette of colours than we can. This illustrates just how important colour is to these animals, and that the ability to see a wide range of bright colours may have been spurred on by colourful plumage that evolved for display purposes in their dinosaur ancestors.

It would be no surprise to find that dinosaurs were as varied and colourful as birds, given they most likely shared the full colour vision of birds. While some mammals are colourful, most tend to be fairly drab, in greys and browns and shades of black and white. This is because, aside from a handful of species (including chimps, orangutans, baboons and humans), they don't see in colour, instead visualising the world in black and white. Mammals also have more of a need to be camouflaged than birds, because they often live on the ground and find it more difficult to flee from predators.

'Birds are brilliantly coloured because they do see in colour, and it's likely that because birds are a kind of dinosaur, the extinct dinosaurs also saw in bright, vivid colour', says Mark Norell. These colours might have helped them recognise other members of their own species, camouflaged them, or been used for defence to dissuade other animals from attacking them (in the same way that some poisonous frogs are thought to use bright colours as a signal that says, 'Don't eat me, I'll make you sick'). Of course, in living birds, some of the most brightly coloured are the males of those species that use colour to woo and court females, such as peacocks and birds of paradise.

Dinosaurs evolved a great array of ornamentation – including crests, frills, horns and spikes – to attract mates, warn off rivals and otherwise communicate. This surely means they used feathers for the same purpose, just as many brightly coloured birds do today. 'Once dinosaurs had acquired feathers for insulation, what could be more natural than to adapt them into display structures?' asks Phil Currie. 'They are lightweight, strong, colourful, and can be shed and replaced.'

Could it even be that the success, diversity and longevity of the dinosaur family is attributable to the bright colours the evolution of feathers afforded them? In combination, the variations in colour and structure can be a powerful tool for creating the differences between isolated populations that allow new species to form. The formation of new species is dependent on there being some sort of barrier to individuals of different populations mating. Feathers are a 'perfect structure to provide such a platform', says Xing Xu.

Though at this stage he says it's just a 'crazy idea', he believes there may even be a way to test the link between feathers and evolutionary success in a group of animals. The first step is to confirm whether the different kinds of feather-, quill- and fluff-like structures seen in fossils of pterosaurs and dinosaurs (such as *Sinosauropteryx, Caudipteryx, Beipiaosaurus, Tianyulong* and *Psittacosaurus*) all share a single evolutionary root, or if they evolved in separate instances. The next step would be to try to reconstruct feather colours and then compare the diversity of brightly coloured groups of dinosaurs and birds with that of groups without such a range of brightly coloured feathers.

Sounds of prehistory

Sitting on my balcony in Sydney early some mornings I am treated to the awesomely loud guttural screeching of a noisy flock of sulphur-crested cockatoos. This garrulous and gregarious group congregates in the trees and on neighbouring balconies. The harsh caws of cockatoos are not pretty, and when I first moved to Australia, I was struck by the contrast to the tuneful twittering of British songbirds. In fact, I've often joked that cockatoos and other noisy Australian birds sound like dinosaurs or pterosaurs — the grating noises they make seem far more fitting for such fierce and imposing animals. But the truth is we don't know very much at all about the noises dinosaurs made.

Until recently, sound was another aspect of dinosaur lives that experts thought we would never know anything about. But a few fascinating anecdotes give us some insight into dinosaur vocalisations. Most birds can make calls today – in fact, birds are the most habitually noisy and melodic members of the animal kingdom – so it's reasonable to assume their dinosaur ancestors employed sound to communicate too.

John Long says we need to take a step back and think about what might have been physiologically possible in dinosaurs. He tells the story of an occasion several years ago when he was producing sound for a clip of dinosaurs to be shown at Museum Victoria in Melbourne. 'The guy at the sound factory immediately mixed a crocodile sound with a lion roaring and came up with this monster-type roar, and he thought that was probably what a *Tyrannosaurus* sounded like', he says. 'And I said: "Nope. You're completely wrong. These things don't have vocal cords." Mammals are the only group that have vocal cords.'

Instead, he says, birds produce sounds by expelling air out of their windpipe and modifying its shape with their mouth, and they sometimes use their tongue to warble too. The sounds of dinosaurs would have been more like those of birds and certainly not like the kind of 'throaty growls' mammals produce. 'The best way to hypothesise about the sounds dinosaurs made is to know the anatomy of their throat, mouth and lungs', Long says. And you never know, with all the fossil dinosaurs coming out of China that have incredibly preserved soft tissues, we may yet get a totally unexpected clue about dinosaur vocalisations.

Jack Horner, palaeontological advisor to Steven Spielberg for the *Jurassic Park* films, has looked to the sounds of birds when creating appropriate soundtracks for museum displays. He agrees that the best approximation we can make for dinosaurs is to take bird vocalisations and slow them down to make them more like the deeper notes that would come from much larger animals. (In case you're wondering, the sounds used in the 1993 film were a whole bunch of manipulated animals noises, from dogs yapping, geese hissing and randy turtles barking, to horses squealing, donkeys yodelling and swans hooting.)

Others scientists have found more direct evidence of the kinds of sounds dinosaurs made. In the early 1980s, David Weishampel, an anatomist at Johns Hopkins University in Baltimore, Maryland, tested an idea that duck-billed hadrosaurs used head crests to make deep vocalisations for communication within herds and family groups, similar to elephants. *Parasaurolophus* were 2.5-tonne plant eaters with large, tubular bony crests that swept back for a metre or more beyond their skulls. The purpose of the crest has been the subject of much debate since the discovery of the species in 1922. Was it used for display, for defence, for sound – or for all three? One strange theory even suggests the crests were used as snorkels.

Weishampel went as far as creating a plastic replica of the crest, which he fitted with a trumpet mouthpiece and played like a wind instrument to simulate at least partially the sounds the animals made. You can find clips of him doing this online, and to my ears it sounds very similar to a didjeridu. He published a paper on this research, arguing that the crest was a good resonator for making powerful low-pitched noises.

In 1996 scientists used supercomputers at the Sandia National Laboratory in Albuquerque, New Mexico (which more typically conducts research into new weapons), to create detailed recon-structions of the shape of *Parasaurolophus* crests and further sim-ulate the kinds of noises that could have been made by forcing air through them. As you would expect, the study revealed that the basic pitch of the note produced was set by the length of the tube. But it also found that the crest had a far more complicated internal structure than had been previously supposed, hinting that the animals might have been able to make a complex and subtle repertoire of calls. Experts at the lab argued that the sound would have been bird-like and might have been used for creating songs as a form of communication.

That research gave some sense of the pitch of vocalisations, showing that young duck-billed dinosaurs would have made higher pitched sounds, and that big adults with large crests on their head would have made deep infrasound bass notes inaudible to human ears. 'Fossil records of the large bones in the dinosaurs' ears compared with corresponding bones in human ears suggest they were able to hear lower frequencies than humans', Carl Diegert of the Sandia lab told reporters.

According to Stephen Brusatte, at the University of Edinburgh in the United Kingdom, another avenue to understanding dinosaur vocalisations has been to use CT scans to look inside the fossils of dinosaur skulls and see the shape of the inner ear and the parts of the brain responsible for processing information from the ear. By comparing the shape of the ears of dinosaurs to those of living animals, researchers have been able to show that animals such as *T. rex* were certainly able to hear very deep sounds.

More recent research has shown that three other hadrosaurs – *Lambeosaurus*, *Corythosaurus* and *Hypacrosaurus* – were also likely to have made deep-pitched calls to communicate with one another, and certainly had ears that would have been able to pick up these low notes.

Phil Senter at Fayetteville State University in North Carolina has studied the evolution of sounds throughout the history of life on earth and shown that birds, and their close relatives crocodiles, make sounds in quite different ways. Birds have a specialised sound-producing structure called a syrinx, which is found on the trachea near the branched opening to the lungs. Crocodiles, on the other hand, use their larynx to make sounds (the vocal cords of mammals are infoldings of membranes across the larynx). The fact that birds and crocodiles evolved such different structures for producing sound perhaps suggests that the common ancestor of these animals didn't produce sounds at all, Senter concluded. His analysis also suggested that not all fossil birds possessed a syrinx, which implies it's an innovation that appeared after birds evolved and therefore was probably not possessed by their dinosaur ancestors.

'The lack of evidence of a syrinx ... will, no doubt, disappoint fans of roaring movie dinosaurs', Senter wrote in a 2008 paper. 'However, lack of ability to vocalise does not necessarily mean that such animals were silent altogether.' Modern reptiles often communicate with one another using non-vocal means of producing sounds, he says. These include hissing, clapping jaws together, grinding lower jaws against upper jaws, rubbing scales together, and using materials in the environment. In addition to their songs, birds also make non-vocal sounds by hissing, bill clapping, stamping and wing beating.

Dinosaurs with feathered wings may therefore have flapped to create noisy displays as some birds do. Even more intriguing is the suggestion from some scientists that the very long tails of sauropods, such as *Diplodocus*, might have been used to create a loud noise like the crack of a whip, perhaps to scare off predators. Cracking a whip creates a shock wave, or sonic boom, that is the result of the tapered tip momentarily exceeding the speed of sound. Research by Phil Currie and computer scientist Nathan Myhrvold predicted that when the long tapering tails of sauropods were flicked from side to side, a wave of energy could propagate along them, gaining momentum and propelling the tip to velocities higher than 1200 kilometres per hour. This is much faster than the speed of sound (and some experts have pointed out this sounds painfully implausible).

We may never know for sure the kinds of sounds dinosaurs made, but by looking at birds and crocodiles we can make some educated guesses. One thing seems fairly certain, though, and that's that dinosaurs didn't produce the kinds of bellows and roars depicted to dramatic and terrifying effect in movies.

Drawing the past

Recreating the sounds and appearance of dinosaurs is an important endeavour. And with all the new discoveries of such well-preserved Chinese dinosaurs with largely complete skeletons, and impressions of soft tissues and feathers, palaeontological illustrators have more to work with than ever before. Information about the colour of feathers is just the latest piece of the puzzle that allows dinosaurs to be reconstructed with previously unexpected accuracy.

The purpose of palaeontology is largely a curiosity exercise, and without artists working alongside them, it's very difficult for scientists to bring fascinating discoveries to wider public attention. Illustrations of feathered dinosaurs can be great works of art in their own right, but they are also one of the most important tools for communicating new discoveries.

It's difficult for the average person to look at a skeleton, or fossil fragments of one, and form an idea of how that animal might have looked in life. If it's an animal we're familiar with, such as a horse or a human, then we already have something to go on, but many dinosaurs and other prehistoric animals are 'wonderfully alien', says Dave Hone, a palaeontologist and blogger at Queen Mary, University of London. 'Illustrations can have huge power to show the scale, size and proportions of dinosaurs, restore missing parts, and give things some real weight and realness that a fossil cannot. That gives people a much better understanding or impression of the issues, and can of course help promote palaeontology generally and encourage interest and support.'

Illustrations of dinosaurs seem so ubiquitous that it's hard to imagine all the work that goes in to them, but rebuilding a living animal from a scattering of fragments requires specialised knowledge of palaeontology and a careful collaboration between artist and scientist. The fact that the first known dinosaurs were initially reconstructed as low-slung creatures with legs out to the sides in a sprawling gait goes to show what a challenge it is to understand extinct anatomy.

'The challenge for the illustrators is to balance the artistic merits of a piece with scientific rigour', says Hone. 'In terms of getting the actual science bit right, there are tons of details: restoring the missing bones; getting the proportions right of all the bones and parts; getting the angles right of the major joints; getting the size and positions of all the muscles right, and making sure they're slack in some places, but bulging and tight in others; creating the details of the nostrils, ears and eyes, not to mention skin scales and feathers. Then there's the environment ... Hence good illustrators take a lot of time, and they know a lot of palaeontology and anatomy, as well as being good artists.'

Illustrating all the new feathered dinosaurs of China has been an especially important task, since the appearance of carnivorous theropods in our imaginations has gone through a revolution from crocodile- or lizard-like creatures to lively animals of an avian persuasion, often sporting brightly coloured plumage.

Luis Rey, one of the world's top dinosaur illustrators, says the main challenge has been to break the comfortable and pervading view of the mass icon. 'Many of us grew up with the image of giant beautiful monsters that became obsolete and died out', he says. 'However, from the 1970s onward we needed to look at them as once-living animals, and also look at their living relatives.'

The renewed study of dinosaur ecology, anatomy, metabolism and posture was met with scepticism at first, as was the later idea that they were the close relatives of birds, he says. 'The last barrier was broken when the first hard evidence of dinosaurs with feathers was found [in 1996]. After that, the icons started to look like lizards with feathery coats, and that also had to be overcome.'

First and foremost, Rey's inspiration for illustration comes from nature. His ideas take into account the fact that animals today have similar patterns for adaptation to natural environments. He also bears in mind that given dinosaurs had colour vision (as birds and reptiles do), they would probably have used skin patterns and colours to blend in, threaten other animals or attract attention. 'Dinosaurs had also feathers, crests and other display items', he says. 'Therefore you can expect that the Mesozoic was indeed a colourful period in life. We are merely beginning to scratch the surface thanks to "Rosetta stone" deposits like the Yixian [Formation of Liaoning] in China.'

Highly skilled palaeoartists such as Luis Rey, Lida Xing, Julius Csotonyi, Gregory S Paul, Peter Schouten, Jan Sovak and Brian Choo have been responsible for bringing feathered dinosaurs to life (as the image section shows). Lida Xing has painted wonderful scenes for *Australian Geographic* of Aussie dinosaurs such as *Minmi*, *Australovenator* and *Diamantinasaurus*. Most excitingly, he created the first ever illustrations of what may be an Aussie tyrannosaur (see image section), as well as a series of seven illustrations of giant marine reptiles, some of which had never been reconstructed before. In these cases we were getting the first exciting glimpse of species that had previously been known only from a smattering of bone fragments.

Research into the sounds and colours of dinosaurs may help artists and filmmakers enliven them in our minds, but what if we could go one step further? Popular culture has often played with the idea of bringing a dinosaur back from the dead. This thrilling idea has seemed nothing more than fiction, but an audacious project conceived in the United States is now looking to the DNA of dinosaurs buried deep within modern birds as a means of attempting just such a thing. Philosophical and ethical issues abound, but reawakening – in a living animal – dinosaur traits that have been asleep for 66 million years would surely be the ultimate tool for stirring the human imagination.

10

Back from the dead

The science and art of reverse engineering a dinosaur.

It's 5 am on a Tuesday in October sometime in the mid-21st century and something astounding is about to happen. Nothing quite like it has happened for tens of millions of years. At a lab in the Midwest of the United States a group of sleep-deprived researchers and other eager observers are gathered around an incubator. They are staring intently at what appears to be a clutch of chicken eggs.

Several of the eggs are trembling and there's an insistent tapping of little creatures struggling to break out. Fluffy feathers start to poke through and cracks appear as the eggs pulse and rock. It takes 90 minutes for the first chick to emerge, but eventually the egg splits down the middle and something covered in wet downy fuzz tumbles out, blinking in the fluorescent light.

It has similarities to a young chicken, but differences too. This animal has a long, reptilian tail and grasping hands where its wings should be, and in place of a beak it has a snout with teeth. The world's first chickenosaurus has emerged, and with it are reborn features of the dinosaurs not seen in a living animal for 66 million years.

Ancient DNA

In Michael Crichton's 1990 novel *Jurassic Park*, dinosaur DNA from the fossil record was used in combination with modern cloning technology to bring long-gone creatures back. Crichton even came up with a seemingly plausible place to find this DNA – inside mosquitoes that had supped on dinosaur blood and become entombed in amber.

But just how plausible is it that we might be able to find dino DNA in the fossil record? And even if we did, would there be enough of it to reconstruct a living animal? DNA is a sensitive molecule that degrades rapidly outside the body. In unusual circumstances – usually very cold and dry conditions – it can be preserved for tens or even hundreds of thousands of years, and work at a number of laboratories specialising in ancient DNA has succeeded in sequencing the genes of extinct species such as mammoths, Tasmanian tigers and Neanderthals. But whereas these species died out within the last 100–50000 years, the dinosaurs have been extinct for 66 *million* years.

A 2012 study of giant moa fossils from New Zealand, led by Morten Allentoft at the University of Copenhagen, with researchers from Murdoch University in Perth, looked at the shelf life of DNA sequences that were long enough to encode any information (and these were still seriously short sequences – just 1 per cent the length of an average human gene). Allentoft's team found that it was vanishingly unlikely that these brief meaningful sequences could survive longer than 1.5 million years – most would have degraded after 158000 years. Studies from the early 1990s claimed to have found sequences of meaningful length in dinosaur fossils and amber, but most experts now agree their samples were probably contaminated with DNA from the researchers themselves or their lab, a common problem in the early days of ancient DNA sequencing. Allentoft's work suggests that even the very shortest sequences of DNA degrade after 6.8 million years.

It therefore seems very unlikely we'll ever find much in the way of dinosaur DNA.

The case is more or less closed, says British dinosaur expert Mike Benton. Even within 100 years, DNA starts to fragment, and so experts have a lot of work to do putting together little scraps of DNA. Anything older than 40 000 years is unlikely to contain useful DNA, he says.

As we saw earlier (see chapter 8), the work of Mary Schweitzer at North Carolina State University and John Asara at Harvard University has shown evidence of blood, tissue and protein in 68-million-year-old fossil bones of *T. rex* and a duck-billed hadrosaur, but this is a long way from finding intact DNA. There's a 'chasm that separates identification of ancient genetic bits from the cloning of a *T. rex*', writes palaeontologist Scott Sampson of the Utah Museum of Natural History in his book *Dinosaur Odyssey: Fossil threads in the web of life*.

As DNA breaks down so rapidly, even if we did have genetic material from the Cretaceous or Jurassic it would be in fragments and missing most of the pieces. 'If this information is not preserved somewhere in the fossil record ... all the technology in the world will be unable to provide a solution', Sampson says:

> Even if you had ... all the genetic material from a dinosaur,
> the tiny fragments would have to be placed in the correct
> order, a gargantuan task equivalent to reconstructing a jigsaw
> puzzle of millions of pieces in which most of the pieces are
> shaped exactly alike. Imagine transforming a book like *War
> and Peace* into a basket full of words and then attempting to
> reconstruct the original text.

In any case, as Jack Horner noted in a 2011 TED talk:

> If you actually had a piece of amber, and it had an insect
> in it, and you drilled into it, and you got something out of

that insect, and you cloned it, and you did it over and over and over again, you'd have a room full of mosquitoes. And probably a whole bunch of trees as well.

Fossils, therefore, probably can't yield the DNA required to reconstruct a dinosaur. But there may be somewhere else to find the information we're seeking – somewhere so obvious that the answer has been sitting right in front of our noses all along. Birds. Birds *are* dinosaurs and they carry the majority of the DNA of their dinosaur forebears within their genes. Could birds therefore be the key to bringing one of their larger and more fearsome relatives back?

The dinosaur species that never were

Jack Horner is a real-life dinosaur hunter at Montana State University's Museum of the Rockies and an impressive character. He overcame a battle with the dyslexia that prevented him from completing an undergraduate degree and became one of the world's foremost dinosaur experts. Then in 1977 he uncovered the first known dinosaur nesting site and first complete specimens of baby dinosaurs (see chapter 8).

Horner is drawn to controversial ideas. One of them is that palaeontologists have greatly overestimated the true number of dinosaur species. Work he has done on dinosaurs collected from the aptly named Hell Creek Formation in the wilds of Montana has drawn interesting parallels between the way birds and dinosaurs develop into adults, and simultaneously sent a whole raft of long-established species into oblivion. Horner has shown that many North American dinosaur 'species' are not species in their own right at all, but merely juveniles of larger animals. Part of the problem, he says, has been that palaeontologists like to name new species, so whenever they find something that looks new and different, they slap a new label on it. But nobody had stopped to wonder where all the juvenile animals were.

It wasn't until 1975 that Peter Dodson, at the University of Pennsylvania, revealed that dinosaurs grew and developed in a similar way to birds, rather than reptiles – with which people had been comparing them until that time. Dodson used the example of the stately and beautiful cassowary, a flightless Australian relative of the emu and ostrich found in the rainforests of northern Queensland and New Guinea. Cassowaries have large casques or crests on their heads, but they grow to about 80 per cent of their full adult size before these appear. If you found an adult and a juvenile side by side in the fossil record you would probably think they were different species.

And this is exactly what Horner says happened with *Dracorex* and *Stygimoloch*, two dinosaurs from Hell Creek in Montana with warty, spiky and bony growths all over their skulls. *Dracorex hogwartsia* was named after the wizarding school in the Harry Potter books and featured on the cover of *National Geographic* magazine in December 2007, but it is actually a juvenile of the dome-headed species *Pachycephalosaurus*, according to a 2009 paper published by Horner and his colleague Mark Goodwin of the University of California, Berkeley.

'Juveniles and adults of these dinosaurs look very, very different … and literally may resemble a different species', Goodwin told reporters:

> But some scientists are confusing morphological differences
> at different growth stages with characteristics that are
> taxonomically important. The result is an inflated number
> of dinosaurs in the late Cretaceous. Early palaeontologists
> recognised the distinction between adults and juveniles,
> but people have lost track of looking at ontogeny – how
> the individual develops – when they discover a new fossil.
> Dinosaurs are not mammals, and they don't grow like
> mammals.

Horner and Goodwin used CT scans and microscopic analysis of bone slices to show that the skulls of *Dracorex* and *Stygimoloch* were 'spongy' and full of air spaces, a trait characteristic of under-developed bone that's still growing rapidly. They argued that the domed head ornaments of *Pachycephalosaurus* – which could have been used for attracting mates as they are in cassowaries, and may have been brightly coloured – began to grow when the dinosaurs reached half their adult size, and continued to change shape and be remodelled throughout life. *Stygimoloch*, with its warty and horny growths, small dome and unfused skull bones, was perhaps a subadult heading towards sexual maturity, while the smaller *Dracorex*, with no dome but abundant spiky growths, was an even younger juvenile.

Though not everyone in the scientific community agrees, Horner's work has provided good evidence that the animal we know as *Triceratops* is actually a smaller form of the fully grown adult we call *Torosaurus* and that a miniaturised tyrannosaur called *Nano-tyrannus* really is just a young *Tyrannosaurus rex*. The key to making this kind of research possible is finding numerous individual fossils of the same species at different stages of development. Luckily, fossils of *Triceratops*, for example, are abundant at Hell Creek, but in more than a century of collecting, museums mostly wanted the largest specimens for their displays, so smaller ones were often ignored. 'We have gone out in the Hell Creek Formation for 11 years doing nothing but collecting absolutely everything we could find, which is the kind of collecting that is required', Horner said at the time of the *Pachycephalosaurus* announcement. 'If you think about *Triceratops*, people had collected for 100 years and still hadn't found any juveniles. And we went out and spent 11 years collecting everything, and we found all kinds of them.'

Horner's work on fossils is fascinating, but as he says himself, there's a point when studying fossils can yield no more informa-tion – and at that point the search for new data has to shift to the genes of living species. This brings us back to the question of whether it could be possible to use modern genes to resurrect a dinosaur of sorts. Horner's latest novel idea and controversial project is the quest to create what he calls the 'chickenosaurus'. He wants to, quite literally, reprogram the genetics and embryonic development of a chicken to hatch out a dinosaur.

Rise of the dino-chicken

Everyone is familiar with the idiom 'rare as hens' teeth' – but the problem is that since 2006, hens' teeth aren't quite so rare after all. An accidental discovery made late one night in the lab by Matthew Harris at the University of Wisconsin revealed a mutant chicken embryo with lumps along its beak that looked suspiciously similar to the teeth seen in developing crocodiles and alligators. This mutant variety of embryo, which is too weak to hatch, was discovered half a century earlier and had been used in studies of embryonic development, but nobody had ever examined its mouth.

'What we discovered were teeth similar to those of crocodiles – not surprising as birds are the closest living relatives of the reptile', said Mark Ferguson co-author of a paper on the find published in the journal *Current Biology*. 'All the pathways to make teeth are preserved which helps us understand how evolutionary changes can be brought about by subtle alterations in developmental biology.'

Following Harris's accidental discovery, the team managed to induce teeth to grow in normal chickens during embryonic development. After an absence of more than 70 million years, birds' teeth were back – and the chance discovery had revealed that birds had never lost the ability to produce teeth, they were just repressing it.

All this got Jack Horner thinking: we've already established that dinosaur DNA is gone from the fossil record, but maybe we can find dinosaur DNA somewhere else. If birds have retained the ability to grow teeth from their dinosaurian ancestors, couldn't they have retained other traits such as tails and hands? Horner argues that we are now much closer to creating a dinosaur and we don't need ancient DNA to achieve it. Birds may look different, but it's cosmetic, he says; underneath they're all dinosaur.

In his book *How to Build a Dinosaur*, co-authored with James Gorman, Horner writes:

> So why not grow a dinosaur? At least that's the thought that
> came to my mind. Leaping over many details, it seemed
> so obvious that if fairly small changes in development,
> which adjusted the timing and concentration of growth and
> signalling, could have led to the evolution of birds from
> nonavian dinosaurs, we could readjust those changes in
> development and get a dinosaur.

The secret to developmentally turning a chicken into a dinosaur lies in the scientific field called evo-devo (shorthand for evolutionary developmental biology). This is the study of how embryos develop in chickens, in humans, and all other animals, how this relates to their genes – and how evolutionary changes can be observed by comparing the embryonic development of different species. Evolutionary traits called atavisms that are throwbacks to the past crop up fairly regularly – even humans briefly have a tail that's reabsorbed in the womb. A series of these atavisms in birds gave Horner the idea of 'reverse engineering' a dinosaur from a chicken. All the information is there, he says, you just have to work out how to switch it back on.

Horner turned to his colleague Hans Larsson, at McGill University in Canada, who studies atavisms by examining the development of bird embryos. He's particularly interested in how birds lost their bony reptilian tail, but also in the transformation of the dinosaur hand and arm into the bird wing. *Velociraptor* had a fearsome-looking hand with a powerful set of claws. This same basic hand structure, albeit with slightly smaller claws, can be seen in fossils of *Archaeopteryx* – but in a modern pigeon or chicken that hand has been transformed into a wing.

Intriguingly, many birds still have claws or spurs on their wings. The most famous example is a species of tropical bird, the hoatzin, of the Amazon and Orinoco river basins in South America. The hoatzin develops claws on its thumb and first digit that emerge from the wings. It hatches from the egg with these claws, then uses them to clamber among the trees until it's able to fly. By adulthood the claws have reduced or perhaps disappeared beneath the feathers. Many other birds still have small claws on one of their digits that are hidden beneath the wings – these include emus, cassowaries and other flightless ratites, and water-fowl such as ducks and geese. These sporadic claws hint that the blueprint for *Archaeopteryx*- or even *Velociraptor*-like clawed hands is still there in the DNA of modern birds.

Although pigeons and chickens may not have claws when they hatch, they do have hands at one stage during their embry-onic development. It's just that later on a gene is switched on that fuses the three fingers or digits together to create a wing. So the key, Horner says, is to find that gene and turn it off to prevent the hand from fusing together. The result would be a chicken with a three-fingered hand like that of *Archaeopteryx*.

Then the goal would be to do the same for tails. Living birds have a rudimentary tail called a pygostyle, but as embryos they have a much longer tail that shrinks during development. Horner wants to turn off the gene that makes the tail disappear. 'What we're trying to do, really, is take our chicken, modify it and make the chickenosaurus', he says. 'It's a cooler looking chicken. But it's just the very basics. I think it's a great way to teach kids about evolutionary biology and developmental biology.'

He believes that, depending on funding, we may be as little as five to 10 years away from creating an animal that wouldn't exactly be a dinosaur but would once more have at least a handful of ancestral dinosaur traits. 'I currently have a team of developmental biologists working on this project', he says. 'The challenges are in finding the right genes to switch on to produce dinosaur-like characteristics … It will reveal at least some of the genetic pathways that were involved in transforming dinosaurs into birds.'

Aside from the fun of seeing dinosaur traits brought back to life, the project could yield fresh insights into development and evolution. Horner even thinks it could prove instructive in curing genetic defects and teaching us how to regenerate damaged bones and spines.

As he says in *How to Build a Dinosaur*, 'One of the benefits of actually hatching a dino-chicken would be that it would be shockingly vivid evidence of the reality of evolution – not a thought experiment but an *Oprah*-ready show-and-tell exhibit. The creature would be its own sound- and vision-bite.'

Frankenstein creatures?

Not everyone is as convinced of the merits of the project, however, nor that chickenosaurus would be anything like a real dinosaur. You do have to wonder whether encouraging the development of a few missing traits, such as tails, teeth and claws, is really creating anything approximating what a dinosaur would have been like, and Horner acknowledges this.

Some critics, particularly religious groups, have decried the project, saying that he is playing God, but his response is that he's not really proposing anything that hasn't already been achieved through selective breeding of dogs, livestock, and most of the fruits, vegetables and grains we eat. It's all genetics, it's just that selective breeding is a slower, less targeted way of achieving the same goals.

Scott Sampson writes in *Dinosaur Odyssey* that reverse engineering a dinosaur may be possible, but he wonders whether it's an endeavour we should be engaging in:

159

With the present astounding rate of advancement in
molecular biology, such genetic monstrosities may be feasible
by the end of this century. Yet it is difficult to imagine that
these Frankenstein creatures would, in any real sense, be
Mesozoic dinosaurs brought back to life.

Benton challenges that perspective, saying it's an interesting
experiment that might reveal suppressed capabilities hidden in
the genomes of birds. 'In the end [Horner] would be the first to
say he's not going to be producing a dinosaur. If you think you can
back-breed from a chicken and end up with a dinosaur, I suspect
not. But at least you can burrow in and look at genetic potential
… People managed to manipulate bird embryos and get them to
produce teeth, and that was a great sensation. Birds haven't had
teeth for maybe 70 or 80 million years, but they still genetically
have that potential to produce teeth, so it must be suppressed.'

The concept of trying to reactivate hidden capabilities is a
smart idea, he says. 'In a way it's using the palaeontology of the
genome, looking to see what's still there – such as whether we
can get a modern bird to produce a long bony tail or claws on
the wings … There's a lot of deep, deep history – hundreds of
millions of years – locked away in the genome.'

Only time will tell whether it might be possible to bring back
a dinosaur, but there have been quite astounding developments
since *Jurassic Park* hit cinema screens in 1993. It's easy to forget
that the cloning technology that was used in the film to breathe
life into dinosaurs was still in its infancy at that time. Dolly the
sheep, the first mammal to be cloned, was not born until 1996. In
1993 people thought it would take an inordinately long time to
sequence a human genome or that it might not be possible at all,
but the first entire human genome sequence was completed only
10 years later. That first genome sequence was a massive effort
involving institutions across the planet and cost an estimated
US$3 billion. Fast forward another decade or so to the present

day and sequencing a genome will soon be available for as little as $1000.

Think of the rapid development of computers, mobile phones and the internet – technology has advanced in ways we can't imagine, making it difficult to predict what might be possible. So you never know, one day in the not-too-distant future, maybe you'll get the chance to (gingerly) pet a little dinosaur too.

11

The survival game

Of the many great dinosaurian lineages, only the birds made it through the mass extinction at the end of the Cretaceous — but nobody is quite sure why.

Sixty-six million years ago, in the last few moments of quiet before the Cretaceous era comes crashing to a close, a herd of duck-billed hadrosaurs is gathered around a small stream at the edge of a forest on a cool and clear night. The reflection of the moon in the water ripples into a thousand pieces as several stoop to drink. Illuminated by the moonlight, small nocturnal dinosaurs festooned in feathers move between the nearby branches; some scramble, others glide or flap, on the hunt for tasty morsels. Sharing the same trees are flocks of birds, most silently perched and sleeping in the darkness. The peace is broken when one of the duck-bills hoots in alarm — a trumpeting sound made with the bony crest on its head. Slowly chewing on ferns and other foliage, the herd peers upward, its attention caught by something searing bright and soundless in the sky.

For many centuries a great celestial ballet has played out across the stage of the solar system. A comet, shaken loose from its original orbit in the distant parts of the sun's dominion beyond Pluto, has been dancing with the planet earth. At first infrequently – but with increasing regularity, during its many circuits of the sun – it has provided a spectacular show in the night sky as the orbits of two satellites wend ever closer. This comet is 10 kilometres in diameter, taller than Mt Everest or larger than the Martian moon Deimos. This harbinger of destruction is travelling at a speed of more than 100 000 kilometres per hour and its energy of motion has the destructive force of 100 million hydrogen bombs.

Tonight the light show will be like nothing dinosaurs have ever witnessed before. As the comet makes its final approach, the earth's gravity draws it into a tight embrace, speeding its passage even further. If this were a film, the movement of this rolling ball of rock and ice might be accompanied by a roaring or rumbling, but space is a vacuum that doesn't transmit sound, so in reality it would have been silent until it hit the planet's atmosphere. As the earth slowly spins, southern North America comes into the comet's sights, specifically the area around what we know today as the Gulf of Mexico and the Yucatán Peninsula. Although the continents have assumed much of their modern configuration by the Late Cretaceous, the climate is warmer and sea levels are higher, so the coastlines look quite different.

It has been a very long time since the earth has endured the impact of an object this size, and the dinosaurs have lived in a relatively stable world for 135 million years. During this time they have adapted, diversified and spread to all its landmasses. All of that is now over, and few animals bigger than 25 kilograms will make it through the next few months alive.

Before the herd of duck-bills even has time to process what it's looking at, there's a blinding flash as the comet punches through the atmosphere and ocean and then plunges beneath the earth's crust, throwing out molten rock from deep below the surface.

The initial crater is 30 kilometres deep and 100 kilometres across, although this rapidly widens and shallows as the rock beneath rebounds. The comet itself is completely vaporised upon impact, and a vast fireball consisting of this vapour, molten rock, dust and other debris is blasted out of the earth's atmosphere and into space. Massive geological activity is triggered across the planet, with earthquakes the size of which we have no concept today. Around the Atlantic itself, gigantic tsunamis, perhaps a kilometre high, race in every direction; fossil evidence of these has been found as far away as Illinois, more than 2000 kilometres to the north.

Smouldering netherworld of the Cretaceous

Everything living within a few hundred kilometres of the impact site – such as the herd of duck-bills – would have been killed almost instantly by the initial fireball, the shockwave of super-heated air, or the rain of molten rock falling back to the planet's surface. But the impacts were felt much more widely than just this central zone of devastation; some theories suggest the atmosphere itself was briefly heated to hundreds of degrees, roasting any animals that couldn't take cover and igniting cataclysmic wildfires that raged across all the continents. According to geologist Walter Alvarez in his book *T. rex and the Crater of Doom*:

> Within hours of the impact, most of Mexico and the United
> States must have been reduced to a desolate wasteland of
> the most appalling, agonising destruction. Where only the
> day before there had been fertile landscapes, full of animals
> and plants of all kinds, now there was a vast, smouldering
> netherworld, mercifully hidden from view by black clouds of
> roiling smoke.

The most distant landmasses from the impact, such as Australia and Antarctica, might have escaped the worst of the initial fire-storm, but tragedy would befall them in the weeks and months ahead. Within days, temperatures across the planet began to plummet as a thick blanket of smoke, dust and soot blocked out the sun's warming rays. Then acid rain, formed from the nitrous oxide and sulfates clogging the atmosphere, began to hammer down on the surface, killing plants and animals and even dissolving rocks. This rain would have been as corrosive as battery acid and its most devastating effect would have been to destroy the shells of small marine organisms. Combined with the atmospheric pollution that blocked out the sun and stopped planktonic algae from photosynthesising, this would have been catastrophic for marine food chains globally, and halted primary productivity for decades. Once the atmosphere cleared and the earth began to warm again, vast quantities of greenhouse gases led to a global warming event that sent temperatures soaring once more for perhaps 1000 years.

'Regionally, there is little doubt that the North American continent would have been absolutely devastated', writes Richard Cowen of the University of California, Davis. 'Globally, even a short-lived catastrophe among land plants and surface plankton at sea would drastically affect normal food chains. Pterosaurs, dinosaurs, and large marine reptiles would have been vulnerable to food shortage.'

Some estimates suggest that as many as 80 per cent of all species went extinct at this time. The dinosaurs, and other large reptiles such as the flying pterosaurs and swimming mosasaurs, weren't the only kinds of animals hit hard. Around 80 per cent of mammals, 83 per cent of snakes and lizards, and even the majority of birds are thought to have gone extinct, although the numbers are unclear because birds preserved as fossils from the late Cretaceous are rare. Birds that disappeared include the enantiornithines – the 'birds with teeth' that were used to make a mockery of OC Marsh in the 1890s (see chapter 3) – and the hesperornithiforms – swimming birds with huge feet that shared the seas with ichthyosaurs and plesiosaurs. Up to 50 per cent of land plants were lost in some areas as well as many plankton species and molluscs, such as the coiled ammonites that make popular fossils today. Whatever combination of events came together at the end of the Cretaceous, we can tell it was a major planet-wide cataclysm, the effects of which were felt in every environment.

The end-Cretaceous mass extinction was devastating, and is the best known because it killed off the dinosaurs, but it's not the biggest extinction event we know about. One earlier catastrophe killed off an even larger proportion of the tree of life. The Permian mass extinction, 252 million years ago, is estimated to have killed 93–97 per cent of all species. The forebears of all living things were in the 3 per cent that survived. The fact we are here at all today is a small miracle considering our ancestors also made it through three other mass extinctions – at the end of the Ordovician (444 million years ago), Devonian (375 million years ago) and Triassic (201 million years ago) periods. You may be wondering why mass extinctions always neatly occur at the end of geological periods. It's because these events are so devastating in the history of life and mark such a change in the species found in fossil deposits that they themselves define how geological time is named and divided. Theories for the causes of these mass extinctions vary from climate change and sea level rise to massive volcanism and extraterrestrial impacts.

One of the more exotic ideas is that there is a 42-million-year cycle in the path of our solar system across the plane of the galaxy. This periodically sees it move above or below of the main disc of the Milky Way making it more likely to be bombarded with comets and cosmic radiation that has a sterilising effect on life. This may have been what killed the dinosaurs at the end of the Cretaceous, the experts claim, but they have little idea how regular the trend is and even whether it really exists or not. In case you're wondering, right now we're hurtling around the galaxy at around 200 kilometres per second and will complete a full circuit in 225 million years – so the last time we were in this spot was during the Triassic, when the first dinosaurs were starting to diversify. When you consider that, on top of that, we're spinning around the axis of the earth at close to 1700 kilometres per hour, and the earth itself is travelling around the sun at 107 000 kilometres per hour, it's enough to make you feel pretty dizzy.

Dino-death whodunnit

A rare element was the clue that led Walter Alvarez and his Nobel Prize–winning physicist father, Luis Alvarez, to the truth about the demise of the dinosaurs. In 1977 Walter was in the central Italian village of Gubbio, 160 kilometres north of Rome, where was he studying magnetism in late Cretaceous rocks. He found something very puzzling – a fine layer of red clay between the limestones of the Cretaceous period and the Paleogene (or Tertiary) period that followed it.

Walter took samples back to the University of California, Berkeley, where he and his father were both based. Luis had the samples chemically analysed for trace elements, which only deepened the mystery. The results showed the red layer had plenty of soot and a level of a metallic element called iridium 30 times that in the surrounding rocks. Silvery-white, brittle and similar to platinum, iridium is very rare on earth but much more common in asteroids and meteorites. Working with the team that had helped them analyse the samples, Walter and Luis went on to find the same layer of iridium in Denmark, New Zealand and numerous other parts of the world. There was only one startling conclusion – something enormous from outer space had smashed into the planet at the end of the Cretaceous period, leaving a fine layer of extraterrestrial iridium all over the globe.

The Alvarezes and their co-workers published their findings in an influential 1980 paper in the journal *Science*, but the idea was greeted with derision from some in the scientific community, who argued that an object large enough to have created a global iridium layer would have left a crater, and none was known from around 66 million years ago. More evidence came in the form rock particles in the same layer (shocked quartz and tektites) that could only have been formed in an exceptionally hot and violent event, such as a nuclear explosion or a meteorite impact.

But the clincher came in 1991, when studies of magnetic and gravitational fields revealed a staggeringly huge crater largely hidden beneath the seafloor at Chicxulub, off the coast of Mexico's Yucatán Peninsula. At 180 kilometres across, this vast feature had been hidden in plain sight. Today, more than 30 years after the first puzzling discovery in Gubbio, the iridium layer has been found at 350 different sites around the world.

While the great majority of experts now agree that an asteroid or comet (the jury is still out on precisely which) did strike the earth around 66 million years ago, not everyone agrees that this was the only reason for the mass extinction, nor on what the effects of the impact would have been.

'There's general agreement that a meteorite hit', says Paul Barrett from the Natural History Museum in London. 'Everyone's happy with that. There's much less agreement on exactly when it hit and what its effects were globally. Some argue it hit a few hundred thousand years before the extinction happened, which might have meant it was a contributing factor but not *the* factor. A number argue that the impact would have had primarily local effects – so a devastating effect on North and South America, but not necessarily a global effect. Also, at this time we know there are massive volcanic eruptions going on in central India, which have been going on for 2 million years sporadically. This would have had significant effects on global climate, throwing sulfur dioxide and carbon dioxide into the atmosphere in vast quantities.'

Gradual changes had been taking place at the end of the Cretaceous anyway. Plate tectonics had caused the continents to drift apart, leading to new ocean circulation and air currents that cooled the planet (and that would soon cause the formerly lush southern landmass of Antarctica to freeze over). There were also

An Australian comet strike

You may not be able to see the Yucatán crater from the surface of the earth, but if you visit the Australian outback you can clearly see the remains of another large comet that struck the planet during the reign of the dinosaurs. Gosse Bluff, a site of stark desert beauty 175 kilometres west of Alice Springs in the Northern Territory, is truly a marvel to behold. A comet hit with such force there it threw up a small mountain range in a circle around the impact site, part of which is still there today. Known as Tnorala to the local Western Arrernte Aboriginal people, the 5-kilometre-wide site is all that remains of a much bigger 20-kilometre-wide crater created 142.5 million years ago when a 600-metre object plummeted to the earth's surface.

major vegetation changes as the new flowering plants started to take over from the conifers.

'You have this perfect storm, if you like', Barrett says. 'Everything is changing in the last few million years of the Cretaceous. Some of it is very rapid, as with the meteorite impact, and some of it is occurring over millions of years, as with the vegetation changes. So all these things are going on … The extinction itself is rapid, but it looks like the global environment is stressed for a long time leading up to this point. We've kind of solved it, because we know all this stuff was going on, but we're never going to know what actually killed the *T. rex* … It will only be answered if we find more fossil sites right at that moment in time. Most of what we know about it currently comes from western North America, as that's where the best record for that time is.'

The unanswered questions are many. A few studies have even hinted that some large dinosaurs might have persisted for hundreds of thousands of years after the main extinction event.

Research from the University of Alberta in Canada suggests that the femur of a hadrosaur from New Mexico is 64.8 million years old. Only time will tell if some dinosaurs really clung onto survival for 700 000 to 1.2 million years after the asteroid hit, or if the difference is just due to discrepancies between dating methods.

One of the biggest unanswered questions, however, is why some animals made it through when others perished. Most particularly, why did some small feathered dinosaurs survive when their much more robust relatives vanished, leaving just a few scattered clues marked in stone for us to attempt to decipher tens of millions of years later?

An idiot's guide to surviving extinction

There appear to be a few golden rules to surviving a mass extinction. Animals that are small, mobile, have lots of offspring, are generalists in habits, range over a wide area and are good at coping with stress are better represented among the survivors of these events than other species. It's clear that birds could tick off many more of these traits than some of their much more massive and specialised dinosaur relatives. Giant sauropods, for example, are likely to have reproduced slowly and taken many years to grow to adult size. They also required vast volumes of specific types of plants, and were much less mobile than small flying animals.

Small animals typically have larger populations and wider genetic diversity. They also reproduce more quickly and can therefore evolve and adapt at a faster rate. Having many offspring creates a wider variety of individuals and makes it more likely that some of them will survive in trying conditions.

Size in particular seems to be an important factor in the survival of birds – no species of any kind larger than 25 kilograms made it past the end of the Cretaceous. 'The obvious explanation for that is that if you are a big animal you need a lot of food', says

Barrett. 'If there's any kind of stress in ecosystems and you have food chains collapsing, then you are not going to do very well. Dinosaurs on average weighed a tonne, so they are all going to be hit very hard ... All birds are small at this time. None would have been bigger than a seagull.'

Being wide-ranging and mobile is something else birds do particularly well today. The Arctic tern, for example, flies an annual round trip of approximately 70 000 kilometres between its Arctic breeding grounds and Antarctic wintering grounds. The oldest known Arctic tern, a bird of 34 ringed in the United States, is therefore likely to have travelled 2.4 million kilometres in its life, further than flying to the moon and back three times. In all, around half of all living bird species migrate in some way each year. And if a species has a distribution across the planet, at least some of those individuals are likely to be somewhere less impacted by the crisis, or to be able to move quickly to such an area. 'Nearly all of the effort of flying is getting off the ground', Barrett says. 'This means that if they live in an area where the resources are really bad, they can just go somewhere else. So these are two big reasons birds might have done well – because they are small and very mobile.'

Being a generalist in habits makes a species more likely to survive trying conditions. Pigeons, which eat whatever they can get their beaks on in cities across the planet, are far more successful, for example, than pandas, which eat only specific types of bamboo in small montane patches of China. 'If you have only one source of food, or need a specific plant to shelter in, then you're doomed if that is affected or taken out', writes Dave Hone. If, however, you can eat pretty much anything, your chance of finding enough to get you through tough times is much greater.

Another factor birds and mammals share is a fluffy covering of insulation in the form of fur and feathers – perhaps this helped shield them from the initial catastrophe or protected them from the severe cold snap that gripped the planet in the weeks and months of darkness after the impact. Some birds and mammals may have been shielded from the heat in burrows or in termite mounds, or sought sanctuary in marine environments and wetlands. Many mammals also hibernate, so may have slept through the cold snap and could have found nuts, seeds and insects in the ground to tide them over until the sun returned.

But this doesn't totally answer the question of how birds, crocodiles and turtles survived the extinction at the end of the Cretaceous while the smallest non-avian dinosaurs (some of which were likely generalists that were common and similar to birds in many ways) didn't. 'The survival of birds is the strangest of all the K–T [Cretaceous–Tertiary] boundary events, if we are to accept the catastrophic scenarios', writes palaeontologist Richard Cowen in his book *History of Life*:

> Smaller dinosaurs overlapped with larger birds in size and in
> ecological roles as terrestrial bipeds. How did birds survive
> while dinosaurs did not? Birds seek food in the open, by
> sight; they are small and warm-blooded, with high metabolic
> rates and small energy stores. Even a sudden storm or a
> slightly severe winter can cause high mortality among bird
> populations.

The surprise is not so much that some birds made it, but that some of the small dinosaurs didn't, argues Hone on his blog. 'Things like dromaeosaurs and troodontids in particular had nearly all of the same characteristics as birds, as far as we can tell, and did similar things, in similar places, and in similar ways. It could of course have just been the luck of the draw; these things do happen. But if you knew in advance some birds would make it, it would have been a decent bet that dromaeosaurs would, too.'

Barrett doesn't have a good explanation for this either. 'Lots of other things the size of small dinosaurs all make it through relatively unscathed', he says. 'There were a number of small-bodied meat eaters and plant eaters that you might have imagined would be immune if body size was the key factor, but for some reason they all go.'

The diets of some birds today are quite malleable, so if they face a period of tough conditions they can switch from eating insects and seeds, or whatever they normally eat, to eating just about anything. The answer might be that small dinosaurs were too specialised in their diet, but we may never have a simple solution.

The benefits of being bird-brained

Other research teams have looked to features of early bird brains, deduced from the shape of the inside of their skulls, to understand why they fared better than other species. These studies have shown that the ancestors of modern birds were not only quicker witted than dinosaurs but also may have had a better sense of smell – both of which would have been useful skills for searching out food in the darkness and chaos that followed the impact.

In a study published in 2009, scientists at the Natural History Museum in London, led by Stig Walsh, put the case that the evolution of a larger and more complex brain may have given the ancestors of modern birds an edge over dinosaurs, pterosaurs and other groups of early birds. 'Birds today are the direct descendants of the Cretaceous extinction survivors, and they went on to become one of the most successful and diverse groups on the planet', Walsh, now at the Natural Museums of Scotland, told journalists. 'There were other flying animals around, such as pterosaurs and older groups of birds, but we've not really known why the ancestors of the birds we see today survived the extinction event and the others did not. It has been a great puzzle for us.'

Among living birds, the most intelligent species – such as New Caledonian crows and many parrots – have more flexible and complex behaviours, such as making use of tools. If the ancestors of modern birds were able to employ their cunning and guile to find food it would have given them a major advantage over other species. And research does indeed suggest that bigger brained birds display behavioural flexibility and are better able to survive in new environments than those with smaller brains – a skill that would have been very useful after the extinction event. 'In the aftermath of the extinction event, life must have been especially challenging', said Walsh. 'Birds that were not able to adapt to rapidly changing environments and food availability did not survive, whereas the flexible behaviour of the large-brained individuals would have allowed them to think their way around the problem.'

Working with palaeontologist Angela Milner, Walsh used CT scans to look at the 55-million-year-old fossil brain cases (as a proxy for the shape of the brain itself) of two early birds that lived in the then warm, tropical conditions of England. They were surprised to find that the brains were likely to have been similar to those of modern birds in the regions that controlled sight, flight and memory, showing that – in organisation and appearance – the brains were essentially modern.

Phoenix rising

Whatever the reason for their survival, the various lineages of modern birds rose phoenix-like from the ashes of the extinction event to diversify, spread and succeed across the planet. They eventually assumed a much wider range of forms and sizes than during the Cretaceous, evolving into creatures as diverse as ducks, emus, penguins, falcons, toucans, kingfishers, flamingos, loons and wrens. They would fly in great flocks that darkened the skies of the Australian outback and populate the icy seas of the newly frozen Antarctica. A million or more of them would stand, at any one time, in the salt lakes of East Africa, while others would seek out and nest in every nook and cranny along the cliffs of the British coastline. Others would develop plumages of brilliant, vibrant colours and swoop screeching between the vines and great hardwoods of the rainforests that grew to dominate the tropics. Though the world was sadly empty of the incredible giants that had dominated it for hundreds of millions of years, the feathered dinosaurs had made it, and their journey into the modern day was well underway.

Relationships of the theropod dinosaurs

Species in parentheses are example species for each group.

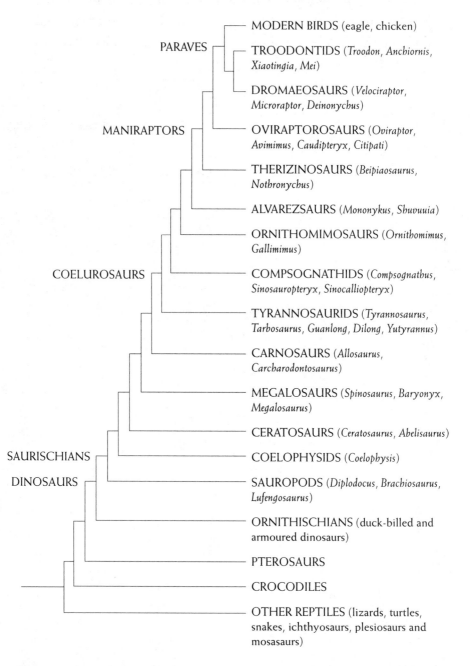

SOURCE: Information from M Benton (2014), *Vertebrate Palaentology*, 4th edn, Hoboken, New Jersey: Wiley-Blackwell; and personal communication with Professor Mike Benton.

An A–Z of feathered dinosaurs

*These are the species for which we have direct evidence of
feathers, either in the form of fossil impressions or features of
bones linked to the presence of feathers in living birds.*

Anchiornis huxleyi
China, Late Jurassic (156–161 million years ago), described 2009
Found in western Liaoning, this chicken-sized species had long,
thickly feathered legs that suggest it was a powerful runner. A
fantastic level of preservation in the fossils has allowed the colour
of its extensive feathers to be reconstructed and show it was black
or grey, with white speckling on the arms, legs and tail, and a
red crest on the head. *Anchiornis* (ANG-kee-OR-niss) means 'near
bird'. This close relative of *Archaeopteryx* was named after early
evolutionary biologist Thomas Henry Huxley, who was the first
to draw a link between dinosaurs and birds.

Anzu wyliei
United States, Late Cretaceous (66 million years ago), described 2014
From the Hell Creek formation of the Dakotas, *Anzu* has been
described by the Smithsonian Institution and Carnegie Museum
scientists who found it as the 'chicken from hell' and a cross
between an emu and a lizard. Only the second feathered dinosaur
ever discovered in the Americas, this omnivorous oviraptorosaur
is from the very end of the dinosaurs' reign. *Anzu* (AHN-zoo) is
the name of a bird-like deity from Sumerian mythology, while
wyliei honours Wylie J Tuttle, the son of a museum donor.

Archaeopteryx lithographica

Germany, Late Jurassic (146–151 million years ago), described 1861
Archaeopteryx is known from 11 late Jurassic fossils, and possibly a twelfth, which is a single feather. It has long been called the first bird, even though it still had reptilian features such as a long bony tail, teeth and three unfused fingers in its wings with claws on the end. It would have been about 45 centimetres long from head to tail. *Archaeopteryx* (ark-ee-OP-ter-iks) means 'ancient wing', while *lithographica* means 'written in stone'. This refers to the shale deposits in which it was found, which were in demand for lithography.

Aurornis xui

China, Late Jurassic (160 million years ago), described 2013
An early relative of *Archaeopteryx*, this pheasant-sized creature pre-dates it by 10 million years. The short length of its feathers, however, suggests it couldn't fly. *Aurornis* (or-OR-nis) *xui* means 'dawn bird named in honour of Xu Xing'.

Avimimus portentosus

Mongolia, Late Cretaceous (71–100 million years ago), described 1981
Discovered by Russian palaeontologist Sergei Kurzanov in the Gobi Desert of Mongolia, this parrot-beaked oviraptorosaur has many bird-like features even though it is not in the group most closely related to birds. It helped build the argument that birds were descended from dinosaurs. Kurzanov argued that the 70-centimetre-tall species had quill knobs on the arm bones, where feathers would have attached, but there was no other evidence at that time for feathered dinosaurs, so the discovery was largely overlooked. *Avimimus* (av-ee-MEE-mus) *portentosus* means 'amazing bird mimic'.

Beipiaosaurus inexpectus

China, Early Cretaceous (120–125 million years ago), described 1999
This was among the first crop of feathered dinosaurs, and was

found in Cretaceous deposits near Beipiao, a city in China's Liaoning Province in the late 1990s. *Beipiaosaurus* was a fabulously weird and shaggy therizinosaur. This 2-metre-long animal would have been covered in fuzzy down-like feathers and had long scythe-like claws. Despite being part of the carnivorous theropod family, its teeth suggest it was a herbivore. *Beipiaosaurus* (bay-pyow-SOR-us) *inexpectus* means 'unexpected lizard from Beipiao'.

Caudipteryx zoui
China, Early Cretaceous (111–125 million years ago), described 1998
This turkey-sized herbivorous oviraptorosaur was the third of the feathered dinosaurs discovered in Liaoning, in 1998. As birds do today, it had a short pygostyle tail, but with a plume of tail feathers that it likely fanned out for display purposes. It had downy feathers across its body and long feathers on its arms. *Caudipteryx* (kaw-DIP-ter-iks) means 'tail feather'. The species was discovered by scientists including Mark Norell and Phil Currie, and was named for Zou Jiahua, a high-ranking Chinese official and a supporter of the palaeontological work at Liaoning. A second, slightly smaller species has been named *Caudipteryx dongi*.

Citipati osmolskae
Mongolia, Late Cretaceous (71–86 million years ago), described 2001
This emu-sized, parrot-beaked oviraptorosaur was the largest of its group until the discovery of *Gigantoraptor* (see below). It has often been found in brooding position on top of clutches of fossilised eggs. First discovered in Mongolia's Gobi Desert in 2001 by a team including Mark Norell of the AMNH, *Citipati* (CHIT-i-puh-tih) is Sanskrit for 'lord of the funeral pyre' and is named for Tibetan Buddhist mythological figures that are often depicted as a pair of dancing skeletons. The species name was given in honour of the Polish palaeontologist Halszka Osmólska.

Concavenator corcovatus
Spain, Early Cretaceous (125–130 million years ago), described 2010
This 6-metre-long, strange-looking animal had two tall vertebrae near the hips, thought to have supported a crest or hump. It has not been found with feathers, but it has quill knobs that suggest feathered forelimbs. *Concavenator* (con-ka-vee-NAY-tor) *corcovatus* means 'hump-backed hunter from Cuenca', after the Spanish province in which its fossil was found.

Conchoraptor gracilis
Mongolia, Late Cretaceous (71–86 million years ago), described 1986
This 1.5-metre-long species was discovered in Mongolia in 1986. Its parrot-like beak is thought to have been adapted to cracking the shells of large marine snails, similar to conch, found in the same fossil beds. Evidence for feathers comes from a 2013 study reappraising its tail stump as a pygostyle. Based on this, experts including Mark Norell and Phil Currie decided it likely had a broad fan of tail feathers for display, similar to *Caudipteryx*, fossils of which have been discovered with feathers. *Conchoraptor* (KONG-ko-RAP-tor) *gracilis* means 'slender conch plunderer'.

Dilong paradoxus
China, Early Cretaceous (120–125 million years ago), described 2004
Discovered by Xu Xing in Liaoning in 2004, this 2-metre predator was a lightly built early relative of *Tyrannosaurus rex*. It was the first tyrannosaurid to be discovered with a downy covering of proto-feathers, and as it is significantly older than *T. rex*, it suggests that all tyrannosaurs may have had feathers, at least as juveniles. *Dilong* (DIE-long) *paradoxus* means 'paradoxical emperor dragon'.

Eosinopteryx brevipenna
China, Late Jurassic (156–161 million years ago), described 2013
This 30-centimetre-long bird-like relative of *Anchiornis* and *Xiaotingia* was found in north-eastern China's Tiaojishan Formation in

Liaoning, in 161-million-year-old deposits. It had short arms, and features of its feathers and toes suggest it scurried about on the ground and hopped between tree trunks rather than flew through the air. *Eosinopteryx* (ee-oh-sin-OP-ter-iks) *brevipenna* means 'short-feathered early Chinese wing'.

Epidexipteryx hui

China, Mid- to Late Jurassic (152–168 million years ago), described 2008
This odd pigeon-sized dinosaur found in China's Inner Mongolia Province had a downy covering for insulation and four long, ribbon-like feathers that emerged from its tail and were probably used for display. Weirder still were *Epidexipteryx's* incredibly long fingers – the third finger was half the length of its entire body. These features suggest it scrambled around in the trees, possibly using the long digits to skewer fat grubs in tree holes and crevices, just as the aye-aye of Madagascar does today. *Epidexipteryx* (EPP-ee-deks-IP-tuh-riks) *hui* means 'Hu's display feather', and the species was named in honour of palaeontologist Hu Yaoming.

Gigantoraptor erlianensis

Mongolia, Late Cretaceous (80–95 million years ago), described 2007
This species was discovered accidentally by Xu Xing while he was shooting a TV show in Inner Mongolia in 2005. The largest member of the parrot-beaked oviraptorosaur group – at 8 metres long and 4 metres tall – it was 35 times the size of the next largest member of the group. It has not been found directly with feather impressions, but is assumed to have had feathers because of its close relationship with known feathered species. It was the largest known animal ever to have had feathers until *Yutyrannus* was described in 2012. Massive nests and eggs likely to have belonged to *Gigantoraptor* have been found in similar Late Cretaceous deposits in Mongolia and China. At 45 centimetres long, the eggs are the largest known dinosaur eggs. *Gigantoraptor* (ji-GAN-to-RAP-tor) *erlianensis* means 'giant plunderer from the Erlian Basin'.

Jianchangosaurus yixianensis

China, Early Cretaceous (125 million years ago), described 2013

A member of the weird therizinosaur group of theropods that returned to herbivory, this 2-metre feathered animal from Liaoning was related to *Beipiaosaurus*. It had a beak and teeth well suited to eating plants. *Jianchangosaurus* (jee-an-CHANG-o-SOR-us) *yixianensis* means 'lizard from the Yixian Formation in Jianchang Province'.

Jinfengopteryx elegans

China, Late Jurassic or early Cretaceous (exact age unknown), described 2005

This was the first troodontid to be discovered with contour feather impressions. Found in the Hebei Province, the 60-centimetre-long fossil had the remains of a grain-based meal inside it, backing up earlier ideas that troodontids may not have been entirely carnivorous. *Jinfengopteryx* (jin-feng-OP-tuh-riks) *elegans* means 'elegant golden phoenix wing'.

Juravenator starki

Germany, Late Jurassic (151–156 million years ago), described 2006

Found in 1998 in similar fine-grained German limestone deposits to *Archaeopteryx*, this species was described by Luis Chiappe. The fossil has scaly skin impressions, but under UV light it also reveals a wispy covering of filaments. This primitive coelurosaur would have reached a length of about 80 centimetres. *Juravenator* (ju-rah-vuh-NAT-tor) means 'hunter from Jura'. The species name honours the Stark family, who own the quarry where the fossil was found.

Microraptor gui

China, Early Cretaceous (110–120 million years ago), described 2003

Hundreds of specimens of several similar species of this four-winged flier have been found in Liaoning. Studies revealed these animals had feathers that were blue–black with an iridescent sheen, similar to a crow, and serrations in their teeth suggest that fish were part of their diet. Three species have been named (*M. gui*, *M. zhaoianus* and *M. hanqingi*), but they may yet be shown to represent variation within a single species. *Microraptor* (MY-crow-RAP-tor) means 'tiny plunderer'. The species was named in honour of Chinese palaeontologist Gu Zhiwei.

Ningyuansaurus wangi
China, Early Cretaceous (125 million years ago), described 2012
This early member of the oviraptorosaur group was found with feather impressions around its tail and gut contents suggesting its diet included seeds. *Ningyuansaurus* (NING-yu-wan-SOR-us) means 'Ningyuan lizard'. Ningyuan is an archaic name for Xing-cheng City, where the single specimen is housed in a museum. The species is named in honour of Wang Qiuwu, who donated the specimen for scientific study.

Nomingia gobiensis
Mongolia, Late Cretaceous (68–71 million years ago), described 2000
This medium-sized, 1.7-metre oviraptorosaur from Mongolia's Gobi Desert was found in 1994 and described by researchers including Canada's Phil Currie in 2000. It has the typical pygo-style tail of the group, which suggests it had a fan of tail feathers it could manoeuvre in a similar way to a peacock or turkey. *Nomingia* (noh-MING-ee-uh) *gobiensis* means 'from the Nomingiin region of the Gobi Desert'.

Ornithomimus edmontonicus
Canada, Late Cretaceous (65–80 million years ago), described 1933
This 3.5-metre-long bipedal runner with long legs and a tooth-less beak had large eyes that suggest it may have been nocturnal. Its diet is contested, but it was probably omnivorous. The genus *Ornithomimus* was first discovered by OC Marsh (see chapter 3) in 1890, who named the partial hind- and forelimb remains he found *O. velox*. The *O. edmontonicus* species was discovered in Alberta

in 1933. A 2012 study led by Darla Zelenitsky at the University of Calgary revealed three Canadian specimens with feather impressions – the first feathered dinosaur fossils discovered in the Americas and some of a handful discovered outside China. One was a juvenile with a downy covering, while two adults had longer pennaceous feathers on their forearms, suggesting they may have been used for mating displays. *Ornithomimus* (or-NITH-o-MEE-mus) means 'bird mimic'. The species is named for the Late Cretaceous Edmontonian faunal stage.

Oviraptor philoceratops

Mongolia, Late Cretaceous (71–86 million years ago), described 1924
Henry Fairfield Osborn turned *Oviraptor* into perhaps the most maligned dinosaur in history when in his 1924 description he accused it of being an egg thief, caught red-handed on top of a tasty nestful of *Protoceratops* eggs. In 1995 more oviraptors were found fanned out over nests, revealing that in fact they were doting parents who had been brooding their own young. Hailing from Mongolia's Gobi Desert, *Oviraptor* was a cassowary-sized bipedal animal that may have had a similar head crest or casque to that species. No specimens have been found with feather impressions, but the species is assumed to have had feathers because of its pygostyle tail. *Oviraptor* (OH-vee-RAP-tor) *philoceratops* means 'egg plunderer, lover of ceratops'.

Pedopenna daohugouensis

China, Late Jurassic (approx. 152–168 million years ago), described 2005
Known only from its hind legs, which had large pennaceous feathers attached to them, *Pedopenna* was a small carnivorous dinosaur. Its discovery by Xu Xing and Zhang Fucheng in 2005 caused some controversy, as this very bird-like species is older than 'first bird' *Archaeopteryx*. *Pedopenna* (PED-oh-PEN-ah) *daohugouensis* means 'feather foot from Daohugou'.

Pelecanimimus polyodon

Spain, Early Cretaceous (125–130 million years ago), described 1994

The fossil of this species shows a long, narrow skull and unusual features that may have acted as attachment points for a throat pouch similar to that of a pelican. An omnivorous species, it would have been around 1.8 metres long. It had 220 tiny teeth, the most of any known theropod. Although they were described as muscle fibres (before the 1996 discovery of feathery *Sinosauropteryx*), the fossil has fine 'integumentary structures' around the neck and arm, which may have been feathers. *Pelecanimimus* (pel-e-KAN-i-MEE-mus) *polyodon* means 'many-toothed pelican mimic'.

Protarchaeopteryx robusta

China, Early Cretaceous (120–125 million years ago), described 1997

This species was similar to *Archaeopteryx*, but with fewer bird-like features (despite appearing 15 million years later in the fossil record) and without the asymmetrical flight feathers necessary for powered flight. Fossils of this turkey-sized animal were found in Liaoning. *Protarchaeopteryx* (pro-tark-ee-OP-ter-iks) means 'first ancient wing'.

Psittacosaurus mongoliensis

China/Mongolia, Early Cretaceous (100–140 million years ago), described 1923

This primitive and largely bipedal ceratopsian with a parrot-like beak was an early member of the group that gave rise to *Triceratops*. Up to 2.5 metres long when fully grown, *Psittacosaurus* was discovered during the AMNH's first foray into Mongolia in 1922. Several similar species of *Psittacosaurus* have been named. A more recent specimen from Liaoning in China (and illegally exported to Germany) has a series of hollow, tube-like, 16-centimetre-long quills along the top of the tail. It is unclear whether these are related to feathers or represent a structure that evolved completely independently. *Psittacosaurus* (SIT-a-ko-SOR-us) *mongoliensis* means 'parrot lizard of Mongolia'.

Rahonavis ostromi

Madagascar, Late Cretaceous (65–71 million years ago), described 1998

Among the first crop of feathered dinosaurs to be discovered in the late 1990s, *Rahonavis* is unique in being from the African island of Madagascar and also hailing from late in the Cretaceous period, relatively close to the time when the dinosaurs became extinct. About the size of a modern raven, this probably flightless species is inferred to have had feathers from the quill knobs on its arm bones. It was found by palaeontologists including Luis Chiappe in 1995, among the bones of a titanosaur, a giant type of sauropod. *Rahona* means both 'menace' and 'cloud' in Malagasy, the indigenous language of Madagascar, so *Rahonavis* (rah-HOON-ah-vis) *ostromi* means 'menacing cloud bird named in honour of John Ostrom'.

Scansoriopteryx heilmanni

China, Late Jurassic (154 million years ago), described 2002

This sparrow-sized tree dweller known from a single specimen from Liaoning is thought to be related to *Archaeopteryx*. Another highly specialised and unusual dinosaur, this feathered animal had an elongated third finger it may have used to skewer insects hidden inside tree hollows, much as the aye-aye of Madagascar does today. *Scansoriopteryx* (scan-SOR-ee-OP-ter-iks) *heilmanni* means 'climbing wing named for Gerhard Heilmann'. Heilmann was an artist and ornithologist who pioneered palaeontological studies of birds in the 1920s.

Sciurumimus albersdoerferi

Germany, Late Jurassic (151–156 million years ago), described 2012

Found in the same fine-grained Bavarian limestone as *Archaeopteryx*, the fossil of this species revealed bristly protofeathers and a puzzling squirrel-style bushy tail. It is one of very few dinosaur fossils from Europe preserved in enough detail to reveal feathers, and one of the most complete predatory dinosaur fossils ever found there. *Sciurumimus* is not in the group of dinosaurs most closely related to birds, so this megalosauroid adds to the evidence that feathers were more widely spread across the dinosaur family tree. *Sciurumimus* (SIGH-oor-uh-MEE-mus) means 'squirrel mimic'. The species is named in honour of Raimund Albersdörfer, a geologist and fossil hunter who made the specimen available for research.

Shuvuuia deserti

Mongolia, Late Cretaceous (71–86 million years ago), described 1998

This odd 1-metre-long Mongolian dinosaur is a member of the alvarezsaurid group, which had short and muscular forelimbs adapted for digging. In *Shuvuuia* these are reduced to one prominent digit, possibly for the purpose of breaking into insect nests. Another unusual trait was a flexible upper jaw it could move independently of its braincase, a feature found in some birds but not known in any other dinosaur. Biochemical analysis of feather-like structures in the type specimen indicated the presence of beta-keratin, the protein found in hair, skin and feathers, and the absence of alpha-keratin, a trait that characterises modern bird feathers. *Shuvuuia* (shu-VOO-ee-a) comes from the Mongolian for 'bird'; *Shuvuuia deserti* therefore means 'desert bird'.

Similicaudipteryx yixianensis

China, Early Cretaceous (120–124 million years ago), described 2008

This oviraptorosaur similar to *Caudipteryx* was deemed to have feathers based on its pygostyle tail when it was found in 2008. Two specimens of different ages were found with feather traces and described in 2010 – differences in the feather arrangements of the adult and the juvenile provided the first evidence that feathered dinosaurs altered their plumage during development in the same way as birds. *Similicaudipteryx* (sim-IL-ee-kaw-DIP-ter-iks) means 'similar to *Caudipteryx*' (which seems rather a lazy name). The species name refers to the Yixian Formation where the fossil was found.

Sinocalliopteryx gigas

China, Early Cretaceous (122–125 million years ago), described 2007

This 2.4-metre-long, 20-kilogram animal was the largest relative of the chicken-sized *Compsognathus*. A 2012 study by Lida Xing, Phil Currie and colleagues showed that *Sinocalliopteryx* was a stealth hunter, based on their analysis of a fossil specimen that appeared to have the remains of the dinosaur *Sinornithosaurus* and the early bird *Confuciusornis* in its gut. *Sinocalliopteryx* (SINE-o-cal-li-OP-ter-iks) means 'beautiful Chinese feather'. The species name, *gigas*, means 'giant'.

Sinornithosaurus millenii

China, Early Cretaceous (120–125 million years ago), described 1999

Around 1 metre long when fully grown, *Sinornithosaurus* was a small predatory dromaeosaur related to *Deinonychus* and *Velociraptor*. The first specimen was discovered by a team including Xu Xing in 1999 and became the fifth feathered species of dinosaur found in western Liaoning. The fossils have downy feathers covering much of the body and more-developed quills on the arms and tail. A 2009 study proposed that the species may have been venomous, injecting poison via fangs in a similar way to a snake, but this has not been widely accepted. A second species, *S. haoiana*, was described in 2004. *Sinornithosaurus* (SINE-or-nith-o-SOR-us) *millenii* means 'millennium Chinese bird lizard'.

Sinosauropteryx prima

China, Early Cretaceous (120–125 million years ago), described 1996

Sinosauropteryx was the dinosaur that started it all – the first feathered species to be discovered – and caused a media sensation when photos of the fossil were brought to a scientific meeting at the AMNH in 1996. This small bipedal predatory dinosaur from Liaoning is a relative of *Compsognathus* and was just 1.3 metres long. Work on fossil evidence of colour-producing structures by Mike Benton's group at the University of Bristol suggests

Sinosauropteryx would have had a russet coat with ginger-and-white stripes around its tail. *Sinosauropteryx* (SINE-o-saw-ROP-ter-iks) *prima* means 'first Chinese reptilian wing'.

Tianyulong confuciusi
China, Late Jurassic (158 million years ago), described 2007
This small ornithischian dinosaur was distantly related to the theropods that gave rise to birds, yet it had a covering of what appear to be feather-like bristles along its neck, back and tail. It is on a primitive branch of the ornithischian group that later gave rise to species such as *Hadrosaurus* and *Triceratops*, which suggests that feathers were widespread across the entire dinosaur family tree. Originally thought to be from the early Cretaceous, it was dated to the late Jurassic by a 2012 study reappraising the rock beds in which it was found. *Tianyulong* (tee-ANN-you-long) *confuciusi* means 'dragon from Tianyu, named in honour of Confucius'.

Velociraptor mongoliensis
Mongolia, Late Cretaceous (71–86 million years ago), described 1924
Velociraptor owes much of its fame to its portrayal as a cunning and collaborative pack hunter in the 1993 movie of *Jurassic Park*, although the animal seen in the movie was closer to the size of *Deinonychus*. Both species of dromaeosaur are distinguished by a large sickle-shaped killing claw on the second toe, which was held aloft from the ground and reserved for slashing at prey. *Velociraptor* was discovered during a 1923 AMNH expedition to Mongolia's Gobi Desert led by Roy Chapman Andrews, and was named in 1924 by Henry Fairfield Osborn. None has yet been found with feathers, but a 2009 study co-authored by Mark Norell pointed to evidence of quill knobs on the bones of the forelimbs, where large feathers would have been attached by ligaments. *Velociraptor* (vuh-LOSS-ee-RAP-tor) *mongoliensis* means 'swift plunderer or thief from Mongolia'.

Xiaotingia zhengi

China, Late Jurassic (156–161 million years ago), described 2011
Another very bird-like species with feathered hind limbs, *Xiao-tingia zhengi* is of similar age to *Anchiornis* and *Archaeopteryx*. When it was described in 2011, it appeared to prove that *Archaeopteryx* was not directly related to the ancestor of birds, but other studies have since disputed this finding. *Xiaotingia* (shyow-TIN-gee-uh) *zhengi* means 'named in honour of Zheng Xiaoting'. This mining magnate founded the Shandong Tianyu Museum of Nature, which houses the largest collection of feathered dinosaurs in the world.

Yixianosaurus longimanus

China, Early Cretaceous (120–125 million years ago), described 2003
This species, known from a single incomplete skeleton, had curiously long hands it may have used for climbing or for snagging prey. *Yixianosaurus* (yee-SHAN-o-SOR-us) *longimanus* means 'long-handed lizard from the Yixian Formation'.

Yutyrannus huali

China, Early Cretaceous (112–125 million years ago), described 2012
Before the announcement of *Yutyrannus* in early 2012 by Xu Xing, most feathered dinosaurs were theropods in the maniraptoran group, and were less than 2 metres long. This bus-sized, 9-metre-long early relative of *T. rex*, found near Beipiao in Liaoning, gave us our first truly scary feathered beastie. The feathers found on the fossil are a downy filament of dinofuzz, perhaps giving this 1.5-tonne predator an incongruously fluffy covering similar to that of a chick. *Yutyrannus* (yoo-ti-RAN-us) *huali* means 'beautiful feathered tyrant'.

References

Quotes and information not specifically referenced here are drawn mainly from a series of interviews conducted between July 2012 and August 2013, and also from earlier research for 'Once were dinosaurs', a feature story published in *Cosmos* magazine in February 2010.

Before we begin

p. xxii The quote from Colin Tudge comes from his 2009 book *The Bird: A natural history of who birds are, where they came from, and how they live*, New York: Crown.

p. xxii The estimate of 1850 total dinosaur species is from Steve C Wang & Peter Dodson (2006), 'Estimating the diversity of dinosaurs', *Proceedings of the National Academy of Sciences*, vol. 103, no. 37, pp. 13601–605, <www.pnas.org/content/103/37/13601.full>.

p. xxiv Information on *Nyasasaurus* is from Sterling J Nesbitt et al. (2012), 'The oldest dinosaur? A Middle Triassic dinosauriform from Tanzania', *Biology Letters*, vol. 9, no. 1, <rsbl.royalsocietypublishing.org/content/9/1/20120949>.

1 The missing link

p. 1 My reconstruction of the living *Archaeopteryx* comes from a variety of sources, namely, J Pickrell (2004), 'Dinosaur-era bird could fly, brain study says', National Geographic News, 4 August, <news.nationalgeographic.com/news/2004/08/0804_040804_archaeopteryx.html>; N Longrich (2006), 'Structure and function of hindlimb feathers in *Archaeopteryx lithographica*', *Paleobiology*, vol. 32, no. 3, pp. 417–31, <www.bioone.org/doi/abs/10.1666/04014.1>; (2013), AM Balanoff et al., 'Evolutionary origins of the avian brain', *Nature*, vol. 501, pp. 93–96, <www.nature.com/nature/journal/v501/n7465/full/nature12424.html>; and (2013), 'X-rays reveal new picture of "dinobird" plumage patterns', media release, University of Manchester, 12 June, <www.manchester.ac.uk/aboutus/news/display/?id=10202>.

p. 5 Pat Shipman's description of Richard Owen is from her 1998 book *Taking Wing: Archaeopteryx and the evolution of bird flight*, New York: Simon & Schuster.

p. 5 Thor Hanson's quote is from his 2012 book *Feathers: The evolution of a natural miracle*, New York: Basic Books.

p. 6 TH Huxley's paper on *Archaeopteryx* and *Compsognathus* was (1868), 'On the animals which are most nearly intermediate between birds and reptiles', *Annals and Magazine of Natural History*, series 4, vol. 2, pp. 66–75.

p. 6 The Huxley quotes about the links between birds and reptiles come from his series of lectures for 'working men' delivered at the Royal School of Mines in 1876 and reprinted in (1902), 'On the evidence as to the origin of existing vertebrate animal', chapter 12 in M. Foster & E. Ray Lankester (eds), *The Scientific Memoirs of Thomas Henry Huxley*, vol. 4, London: Macmillan, pp. 163–87.

p. 7   Huxley's comments about the 'half-hatched chicken' appeared in (1870), 'Further evidence of the affinity between the dinosaurian reptiles and birds', *Proceedings of the Royal Geological Society*, vol. 26, no. 1, pp. 12–31.

p. 8   John Ostrom's 1975 article is 'Archaeopteryx', *Discovery*, vol. 11, no. 1, pp. 15–23.

p. 9   Kevin Padian's comment is from Robert Sanders (2007), 'Agonized pose tells of dinosaur death throes', media release, University of California, Berkeley, 6 June, <www.berkeley.edu/news/media/releases/2007/06/06_deaththroes.shtml>.

p. 11   Achim Reisdorf and Michael Wuttke's 2012 paper on neck ligaments is 'Re-evaluating Moodie's opisthotonic-posture hypothesis in fossil vertebrates part I: Reptiles – The taphonomy of the bipedal dinosaurs *Compsognathus longipes* and *Juravenator starki* from the Solnhofen Archipelago (Jurassic, Germany)', *Palaeobiodiversity and Palaeoenvironments*, vol. 92, no. 1, pp. 119–68, <link.springer.com/article/10.1007/s12549-011-0068-y>.

p. 13   Alan Turner and Mark Norell's comments about quill knobs on *Velociraptor* were reported in (2007), '*Velociraptor* had feathers', media release, American Museum of Natural History, 20 September, <www.amnh.org/science/papers/velociraptor_feathers.php>.

p. 13   The *New York Times* article is John Noble Wilford (2005), 'John H. Ostrom, influential paleontologist, is dead at 77', *New York Times*, 21 July, <www.nytimes.com/2005/07/21/nyregion/21OSTROM.html>.

p. 15   John Long's comments on *Deinonychus* and *Velociraptor* come from his 2008 book with Peter Schouten, *Feathered Dinosaurs: The origin of birds*, Melbourne: CSIRO Publishing.

2   A feathered revolution begins

p. 18   The quotes from Li Yinfang come from Dan Chinoy (2009), 'Walking among dinosaurs', *China Daily*, 24 November, <www.chinadaily.com.cn/cndy/2009-11/24/content_9027450.htm>.

p. 21   The *New York Times* story on *Sinosauropteryx* is Malcolm W Browne (1997), 'In China, a spectacular trove of dinosaur fossils is found', *New York Times*, 25 April, <www.nytimes.com/1997/04/25/us/in-china-a-spectacular-trove-of-dinosaur-fossils-is-found.html>.

p. 22   The *Nature* paper by Lawrence M Witmer was (2009), 'Dinosaurs: fuzzy origins for feathers', *Nature*, vol. 458, pp. 293–95, <www.nature.com/nature/journal/v458/n7236/full/458293a.html>.

p. 26   The descriptions of *Protarchaeopteryx* and *Caudipteryx* appear in Philip J Currie, Mark A Norell & Ji Shuan (1998), 'Two feathered dinosaurs from northeastern China', *Nature*, vol. 393, pp. 753–61, <www.nature.com/nature/journal/v393/n6687/full/393753a0.html>.

p. 28   The quotes from Alan Feduccia appear in Pat Shipman's 1998 book *Taking Wing: Archaeopteryx and the evolution of bird flight*, New York: Simon & Schuster.

3 The dinosaur hunters

p. 30   The quote from Xu Xing to the film crew at the *Sonidosaurus* dig site are from David Cyranoski (2007), 'Giant bird-like dinosaur found', *Nature News*, 13 June, <www.nature.com/news/2007/070611/full/news070611-9.html>.

p. 40   The description by Url Lanham comes from his 2012 book *The Bone Hunters: The heroic age of paleontology in the American West*, New York: Dover Publications.

p. 40   Charles Darwin's letter to Marsh can be read at 'A Yale tale: Darwin's letter to O.C. Marsh', Peabody Museum of Natural History, Yale University, <archive.peabody.yale.edu/exhibits/fossils/history>.

p. 40   The James Penick quote comes from his 1971 paper 'Professor Cope vs. Professor Marsh', *American Heritage*, vol. 22, no. 5, August, pp. 5–13.

p. 42   Tom Huntington's *American History* article is (1998), 'The great feud', *American History*, August, <www.historynet.com/the-great-feud-august-98-american-history-feature.htm>.

p. 42   OC Marsh's description of *Pterodactylus oweni* is (1871), 'Note on a new and gigantic species of Pterodactyle', *American Journal of Science*, series 3, vol. 1, no. 6, pp. 447–59.

p. 45   The Mark Jaffe quote comes from his 2000 book *The Gilded Dinosaur: The fossil war between E.D. Cope and O.C. Marsh and the rise of American science*, New York: Crown.

p. 46   The 'smash-and-grab' comment is from David Rains Wallace (1999), *The Bonehunters' Revenge: Dinosaurs, greed, and the greatest scientific feud of the gilded age*, Boston: Houghton Mifflin.

p. 46   The comments from Mike Benton concerning the 'bone wars' come from his 2008 paper 'Fossil quality and naming dinosaurs', *Biology Letters*, vol. 4, pp. 729–32, <rsbl.royalsocietypublishing.org/content/4/6/729.full>.

4 From dinosaur to bird

p. 47   Mark Norell and Xu Xing's 2004 description of *Mei long* is 'A new troodontid dinosaur from China with avian-like sleeping posture', *Nature*, vol. 431, pp. 838–41, <www.nature.com/nature/journal/v431/n7010/abs/nature02898.html>.

p. 48   The second *Mei long* specimen is reported in Chunling Gao et al. (2012), 'A second soundly sleeping dragon: new anatomical details of the Chinese troodontid *Mei long* with implications for phylogeny and taphonomy', *PLoS ONE*, vol. 7, no. 9, e45203, <www.plosone.org/article/info%3Adoi%2F10.1371%2Fjournal.pone.0045203>.

p. 48   Colin Tudge's comment on *Mei long* comes from his 2009 book *The Bird: A natural history of who birds are, where they came from, and how they live*, New York: Crown.

p. 50   The analysis of pneumatisation in *Aerosteon riocoloradensis* is reported in Paul Sereno et al. (2008), 'Evidence for avian intrathoracic air sacs in a new predatory dinosaur from Argentina', *PLoS ONE*, vol. 3, no. 9, e3303, <www.plosone.org/article/info%3Adoi%2F10.1371%2Fjournal.pone.0003303>.

p. 51   Uncinate processes are reported in *Velociraptor* and *Archaeopteryx* in Jonathan R Codd et al. (2008), 'Avian-like breathing mechanics in maniraptoran dinosaurs', *Proceedings of the Royal Society B*, vol. 275, no. 1631, pp. 157–61,

<rspb.royalsocietypublishing.org/content/275/1631/157>.

p. 52   The 2001 paper by Kevin Padian and other experts was 'Dinosaurian growth rates and bird origins', *Nature*, vol. 412, pp. 405–408, <www.nature. com/nature/journal/v412/n6845/full/412405a0.html>.

p. 52   The find of baby dinosaur bones in Lufeng is announced in Lanna Crucefix (2013), 'Oldest dinosaur bonebed reveals embryo development', media release, University of Toronto, 10 April, <news.utoronto.ca/oldest-dinosaur-bonebed-reveals-embryo-development>.

p. 53   The Michael Novacek quote regarding warm-bloodedness comes from the 2012 AMNH video 'Dinosaurs explained', American Museum of Natural History, <www.amnh.org/explore/amnh.tv/%28watch%29/dinosaurs-explained/were-dinosaurs-warm-blooded/%28p%29/1>.

p. 57   Clark and Xu's study of *Limusaurus* is Xu Xing et al. (2009), 'A Jurassic ceratosaur from China helps clarify avian digital homologies', *Nature*, vol. 459, pp. 940–44, <www.nature.com/nature/journal/v459/n7249/full/nature08124.html>.

p. 57   Xu's comments to *Nature* about *Limusaurus* are in Matt Kaplan (2009), 'Dinosaur's digits show how birds got wings', *Nature News*, 17 June, <www.nature.com/news/2009/090617/full/news.2009.577.html>.

p. 58   The 2009 study of hummingbird genome size is T Ryan Gregory et al. (2009), 'The smallest avian genomes are found in hummingbirds', *Proceedings of the Royal Society B*, vol. 276, no. 1674, pp. 3753–57, <rspb. royalsocietypublishing.org/content/early/2009/08/05/rspb.2009.1004.full>.

p. 60   The study of disease evidence in *T. rex* jaws is Ewan DS Wolff et al. (2009), 'Common avian infection plagued the tyrant dinosaurs', *PLoS ONE*, vol. 4, no. 9, e7288, <www.plosone.org/article/info:doi/10.1371/journal. pone.0007288>.

p. 61   The information on the discovery of Sue comes from (n.d.), 'Sue at the Field Museum', Field Museum, <archive.fieldmuseum.org/sue/ - index>.

p. 61   The *New York Times* quote appeared in Malcolm W Browne (1996), 'Fetching T. rex fossil may fetch $1 million plus, expert says, *New York Times*, 16 November, <www.nytimes.com/1996/11/16/us/fetching-t-rex-fossil-may-fetch-1-million-plus-expert-says.html>.

p. 62   The study suggesting dinosaur lice is Vincent S Smith (2011), 'Multiple lineages of lice pass through the K–Pg boundary', *Biology Letters*, doi: rsbl.2011.0105v1-rsbl20110105, <rsbl.royalsocietypublishing.org/content/early/2011/04/01/rsbl.2011.0105.full>.

p. 62   The announcement of a *Sinocalliopteryx* fossil with remains of other animals in its gut is Lida Xing et al. (2012), 'Abdominal contents from two large Early Cretaceous compsognathids (Dinosauria: Theropoda) demonstrate feeding on confuciusornithids and dromaeosaurids', *PLoS ONE*, vol. 7, no. 8, e44012, <www.plosone.org/article/info:doi/10.1371/journal. pone.0044012>.

p. 64   The *Evolution* study of *Microraptor* is Lida Xing et al. (2013), 'Piscivory in the feathered dinosaur *Microraptor*', *Evolution*, vol. 67, no. 8, pp. 2441–45, <onlinelibrary.wiley.com/doi/10.1111/evo.12119/full>.

p. 64   Eric Snively is quoted in (2013), 'Allosaurus fed more like a falcon than a crocodile, new study finds', media release, Ohio University, 21 May,

<www.ohio.edu/compass/stories/12-13/5/allosaurus-story.cfm>.

p. 65   The suggestion that *Sinornithosaurus* may have been venomous was reported
in Enpu Gong et al. (2010), 'The birdlike raptor *Sinornithosaurus* was
venomous', *Proceedings of the National Academy of Sciences*, vol. 107, no. 2,
pp. 766–68, <www.pnas.org/content/107/2/766.full>.

p. 65   The *Paläontologische Zeitschrift* article refuting the suggestion that
Sinornithosaurus was venomous is Federico A Gianechini et al. (2010), 'A
reassessment of the purported venom delivery system of the bird-like
raptor *Sinornithosaurus*', *Paläontologische Zeitschrift*, vol. 85, no. 1, pp. 103–107,
<link.springer.com/article/10.1007/s12542-010-0074-9>.

5 Fake fossils

p. 66   The notorious *National Geographic* article on *Archaeoraptor liaoningensis* was
Christopher P Sloan (1999), 'Feathers for *T. rex*?', *National Geographic*, vol.
196, no. 5, pp. 98–107.

p. 67   The quotes from Mark Norell come from his 2005 book with Mick Ellison,
Unearthing the Dragon: The great feathered dinosaur discovery, New York: Pi Press.

p. 68   Lawrence M Witmer's description of Liaoning and its fossils appears in his
2009 paper 'Dinosaurs: fuzzy origins for feathers', *Nature*, vol. 458,
pp. 293–95, <www.nature. com/nature/journal/v458/n7236/full/458293a.
html>.

p. 72   Xiaoming Wang's 2013 *PNAS* opinion piece was 'Mortgaging the future of
Chinese paleontology', *Proceedings of the National Academy of Sciences*, vol. 110,
no. 9, p. 3201, <www.pnas.org/content/110/9/3201.full>.

p. 73   The investigative report published in *Science* is Richard Stone (2010),
'Altering the past: China's faked fossils problem', *Science*, vol. 330, no. 6012,
pp. 1740–41, <www.sciencemag.org/content/330/6012/1740.short>.

p. 75   The *China Daily* report about reclaimed fossils is Wang Qian (2010),
'Clampdown on fossil smuggling', *China Daily*, 25 November,
<www.chinadaily.com.cn/china/2010-11/25/content_11605554.htm>.

p. 75   Mark Norell's letter regarding the Mongolian *Tarbosaurus bataar* fossil is
reproduced at 'Provenance of Mongolia Tarbosaurus being auctioned',
Dinosaur Mailing List, Cleveland Museum of Natural History, 18 May,
<dml.cmnh.org/2012May/msg00133.html>.

p. 76   Mongolian president Elbegdorj Tsakhia's statement is quoted in Larry
Neumeister (2012), 'Government in NY seizes "Ty" the dinosaur',
Associated Press, 22 June, <bigstory.ap.org/article/government-ny-seizes-
ty-dinosaur>.

p. 77   Phil Currie's 2012 column was 'Fossil bounty hunters' days may be
numbered', *New Scientist*, 19 June, <www.newscientist.com/article/
mg21428690.200-fossil-bounty-hunters-days-may-be-numbered.html>.

p. 78   The 2014 seizure of a *Tarbosaurus* skull in Wyoming is reported in W Parry
(2014), 'Bad to the bone: dealer pleads guilty in fossil smuggling scheme',
Livescience, 7 January, <www.livescience.com/42374-dinosaur-fossil-
smuggling.html>.

p. 79   The report on the internal *National Geographic* investigation appeared as
Lewis M Simons (2000), 'Archaeoraptor fossil trail', *National Geographic*,
vol.198, no. 4, pp. 128–32.

p. 80   The findings regarding the forgery of *Archaeoraptor* from 88 different pieces appear in Timothy Rowe et al. (2001), 'Forensic palaeontology: the *Archaeoraptor* forgery', *Nature*, vol. 410, pp. 539–40, <www.nature.com/nature/journal/v410/n6828/full/410539b0.html>.

p. 81   The discovery of *Microraptor* is reported in Xu Xing et al. (2000), 'The smallest known non-avian theropod dinosaur', *Nature*, vol. 408, pp. 705–708, <www.nature.com/nature/journal/v408/n6813/full/408705a0.html>.

p. 81   The discovery of *Yanornis* is reported in Zhou Zhonghe et al. (2002), 'Archaeoraptor's better half', *Nature*, vol. 420, p. 285, <www.nature.com/nature/journal/v420/n6913/abs/420285a.html>.

6 The evolution of feathers

p. 83   The description of *Epidexipteryx hui* appeared in Zhang Fucheng et al. (2008), 'A bizarre Jurassic maniraptoran from China with elongate ribbon-like feathers', *Nature*, vol. 455, pp. 1105–108, <www.nature.com/nature/journal/v455/n7216/full/nature07447.html>.

p. 84   John Long's speculations as to the functions of feathers appear in his 2008 book with Peter Schouten, *Feathered Dinosaurs: The origin of birds*, Melbourne: CSIRO Publishing.

p. 85   Kevin Padian's discussion of *Sinosauropteryx* is (1998), 'When is a bird not a bird?', *Nature*, vol. 393, pp. 729–30, <www.nature.com/nature/journal/v393/n6687/full/393729a0.html>.

p. 88   The *Similicaudipteryx* research reported in *Nature* in 2010 is Xu Xing et al. (2010), 'Exceptional dinosaur fossils show ontogenetic development of early feathers', *Nature*, vol. 464, pp. 1338–41, <www.nature.com/nature/journal/v464/n7293/full/nature08965.html>.

p. 88   Xu Xing's comment regarding baby *Similicaudipteryx* flight feathers appeared in Janet Fang (2010), 'Dinosaurs outgrow their baby feathers', *Nature News*, 28 April, <www.nature.com/news/2010/100428/full/news.2010.208.html>.

p. 89   Darla Zelenitsky's research regarding moulting in *Ornithomimus* was reported in Darla K Zelenitsky et al. (2012), 'Feathered non-avian dinosaurs from North America provide insight into wing origins', *Science*, vol. 338, no. 6106, pp. 510–14, <www.sciencemag.org/content/338/6106/510>.

p. 89   The quote from Darla Zelenitsky regarding the first feathered dinosaurs from the western hemisphere comes from (2012), 'Canadian researchers discover fossils of first feathered dinosaurs from North America', media release, University of Calgary, 25 October, <www.ucalgary.ca/news/releases/october2012/first_feathered_dinosaurs>.

p. 91   The discovery of *Yutyrannus huali* was announced in Xu Xing et al. (2012), 'A gigantic feathered dinosaur from the Lower Cretaceous of China', *Nature*, vol. 484, pp. 92–95, <www.nature.com/nature/journal/v484/n7392/full/nature10906.html>.

p. 92   Colin Tudge's contention that adult *T. rex* were feathered appears in his 2009 book *The Bird: A natural history of who birds are, where they came from, and how they live*, New York: Crown.

p. 93   The discovery of *Sciurumimus* was reported in Oliver WM Rauhut et al. (2012), 'Exceptionally preserved juvenile megalosauroid theropod dinosaur with filamentous integument from the Late Jurassic of Germany', *Proceedings of the National Academy of Sciences*, vol. 109, no. 29, pp. 11746–51, <www.pnas.org/content/109/29/11746>.

p. 93   Oliver Rauhut's comments on *Sciurumimus* appear in (2012), 'Newly discovered dinosaur implies greater prevalence of feathers', media release, American Museum of Natural History, 2 July, <www.amnh.org/science/papers/feathers.php>.

p. 94   The study published in 2014 by Paul Barrett and David Evans of whether feathers occurred in other ornithischians is previewed in Matt Kaplan (2013), 'Feathers were the exception rather than the rule for dinosaurs', *Nature News*, 27 December, <www.nature.com/news/feathers-were-the-exception-rather-than-the-rule-for-dinosaurs-1.14379>.

p. 95   Dave Hone's comments about dinosaurs and feathers appeared in his blog in 2013, 'Feathered everything: just how many dinosaurs had feathers?', 'Lost worlds', *The Guardian*, 10 June, <www.theguardian.com/science/lost-worlds/2013/jun/10/dinosaurs-fossils>.

p. 97   Alfred Russel Wallace's comments on the origin of feathers appeared in Harold Begbie (1910), 'New thoughts on evolution: views of Professor Alfred Russel Wallace, O.M., F.R.S.', *Daily Chronicle*, 3 and 4 November, see <people.wku.edu/charles.smith/wallace/S746.htm>.

p. 98   The story of Richard Prum's hearing loss is retold in C Shufro (2011), 'The bird-filled world of Richard Prum', *Yale Alumni Magazine*, vol. 75, no. 2, November/December, <www.yalealumnimagazine.com/articles/3318?page=1>.

p. 100   The *Scientific American* article by Richard Prum and Alan Brush is (2003), 'Which came first, the feather or the bird?', *Scientific American*, March.

p. 102   The paper describing the feathers found in Canadian amber is Ryan C McKellar et al. (2011), 'A diverse assemblage of Late Cretaceous dinosaur and bird feathers from Canadian amber', *Science*, vol. 333, no. 6049, pp. 1619–22, <www.sciencemag.org/content/333/6049/1619>.

p. 103   Mark Norell's commentary on the Canadian amber is (2011), 'Fossilized feathers', *Science*, vol. 333, no. 6049, pp. 1590–91, <www.sciencemag.org/content/333/6049/1590>.

7   The struggle to the skies

p. 109   Alan Turner's findings on dinosaur downsizing are outlined in (2012), 'Dinosaur fossil showing early signs of miniaturization necessary for flight found by AMNH paleontologists', media release, American Museum of Natural History, 6 September, <www.amnh.org/science/papers/mahakala.php>.

p. 109   The 2007 *Science* paper from Alan Turner on dinosaur size is Alan H Turner et al. (2007), 'A basal dromaeosaurid and size evolution preceding avian flight', *Science*, vol. 317, no. 5843, pp. 1378–81, <www.sciencemag.org/content/317/5843/1378.full?sid=8b3b43c1-aeea-4e4e-9bc4-e45a61d5e00c>.

p. 110   The theory of 'stability flapping' appears in Denver W Fowler et al. (2011), 'The predatory ecology of *Deinonychus* and the origin of flapping in birds', *PLoS ONE*, vol. 6, no. 12, e28964, <www.plosone.org/article/

info%3Adoi%2F10.1371%2Fjournal.pone.0028964>.

p. 111   Thor Hanson's comments on the evolution of flight appear in his 2012 book *Feathers: The evolution of a natural miracle*, New York: Basic Books.

p. 113   The description of *Anchiornis huxleyi* appears in Hu Dongyu et al. (2009), 'A pre-*Archaeopteryx* troodontid theropod from China with long feathers on the metatarsus', *Nature*, vol. 461, pp. 640–43, <www.nature.com/nature/journal/v461/n7264/full/nature08322.html>.

p. 113   The *Science* paper about hind-limb feathers is Xiaoting Zheng et al. (2013), 'Hind wings in basal birds and the evolution of leg feathers', *Science*, vol. 339, no. 6125, pp. 1309–12, <www.sciencemag.org/content/339/6125/1309>.

p. 115   The Richard Dawkins quote is from his 2006 book *The God Delusion*, Boston: Houghton Mifflin.

p. 115   Ken Dial's comments on the evolution of flight appear in Cary Shimek (2008), 'Unlocking the secrets of flight: UM releases new theory of bird evolution', *Vision*, University of Montana, <www2.umt.edu/urelations/vision/2008/Secrets of Flight.html>.

p. 116   Nina Schaller's research on ostrich running mechanics is outlined in (2010), 'Feathered friends: ostriches provide clues to dinosaur movement', media release, Society for Experimental Biology, 2 July, <www.sciencedaily.com/releases/2010/06/100630213614.htm>.

8   Sex for *T. rex*

p. 120   The paper on *Oviraptor* brooding its young was Mark A Norell (1995), 'A nesting dinosaur', *Nature*, vol. 378, pp. 774–76, <www.nature.com/nature/journal/v378/n6559/pdf/378774a0.pdf>.

p. 120   Mark Norell and David Weishampel's comments to the *Philadelphia Inquirer* were reported in Mark Jaffe (1995), 'Fossil shows that dinosaur selflessly tended its nest', *Philadelphia Inquirer*, 21 December, <articles.philly.com/1995-12-21/news/25668282_1_oviraptor-fossilized-eggs-mark-norell>.

p. 121   Mary Schweitzer and Jack Horner's paper on medullary bone in dinosaurs is Mary H Schweitzer et al. (2005), Gender-specific reproductive tissue in ratites and *Tyrannosaurus rex*, *Science*, vol. 308, no. 5727, pp. 1456–60, <www.sciencemag.org/content/308/5727/1456.short>.

p. 122   The eggs found by researchers at the Autonomous University of Barcelona were reported in Nieves López-Martínez & Enric Vicens (2012), 'A new peculiar dinosaur egg, *Sankofa pyrenaica* oogen. nov. oosp. nov. from the Upper Cretaceous coastal deposits of the Aren Formation, south-central Pyrenees, Lleida, Catalonia, Spain', *Palaeontology*, volume 55, no. 2, pp. 325–39, <onlinelibrary.wiley.com/doi/10.1111/j.1475-4983.2011.01114.x/full>.

p. 128   Paul Else's comments on possible dinosaur lactation appear in his 2013 paper, 'Dinosaur lactation?', *Journal of Experimental Biology*, vol. 216, pp. 347–51, <jeb.biologists.org/content/216/3/347>.

p. 130   John Long's *Dawn of the Deed: The prehistoric origins of sex* was published in Chicago by the University of Chicago Press in 2012. The quote about eversible penises also comes from this book.

p. 131 Kenneth Carpenter's comments on sexing *T. rex* appear in his 1999 book *Eggs, Nests, and Baby Dinosaurs: A look at dinosaur reproduction*, Bloomington and Indianapolis: Indiana University Press.

9 Colouring in the dinosaurs

p. 133 The research on the colours of ichthyosaurs and mosasaurs is presented in Lindgren, J et al. (2014), 'Skin pigmentation provides evidence of convergent melanism in extinct marine reptiles', *Nature*, doi:10.1038/nature12899, <www.nature.com/nature/journal/vaop/ncurrent/full/nature12899.html>.

p. 134 The quote from Lindgren was reported in (2014), 'Unique fossil pigments found', media release, Lund University, 9 January, <www.lunduniversity.lu.se/o.o.i.s?id=24890&news_item=6120>.

p. 135 The *Daily Mail* article on *Sinosauropteryx* is David Derbyshire (2010), 'Oh no, it's ginger-saurus! For first time scientists uncover colour of dinosaur and it was … a red-head', *Daily Mail*, 29 January, <www.dailymail.co.uk/sciencetech/article-1246535/Dinosaurs-colour-discovered-time--GINGER.html>.

p. 138 Maria McNamara's comments to *Nature* appeared in Ed Yong (2013), 'Dust-up over dinosaurs' true colours', *Nature News*, 27 March, <www.nature.com/news/dust-up-over-dinosaurs-true-colours-1.12674>.

p. 141 Mark Norell talks about birds and colour in the 2012 AMNH video 'Dinosaurs explained', American Museum of Natural History, <www.amnh.org/explore/amnh.tv/%28watch%29/dinosaurs-explained/were-dinosaurs-warm-blooded/%28p%29/1>.

p. 141 Phil Currie talks about feathers as display structures in his 2000 article 'Feathered dinosaurs', in Gregory S. Paul (ed.), *The Scientific American Book of Dinosaurs*, New York: Byron Preiss Visual Publications and Scientific American, pp. 183–89.

p. 145 Carl Diegert's comments on dinosaur ear bones were reported in (1997), 'Scientists use digital paleontology to produce voice of *Parasaurolophus* dinosaur', media release, Sandia National Laboratories, 5 December, see <www.sciencedaily.com/releases/1997/12/971210064258.htm>.

p. 146 Phil Senter's paper discussing dinosaurs' lack of a syrinx was (2008), 'Voices of the past: a review of Paleozoic and Mesozoic animal sound', *Historical Biology*, vol. 20, no. 4, pp. 255–87, <www.tandfonline.com/doi/abs/10.1080/08912960903033327 - .UY8uwcqLHUE>.

p. 146 The use of dinosaur tails to make whip-like noises is suggested in Nathan P Myhrvold & Philip J Currie (1997), 'Supersonic sauropods? Tail dynamics in the diplodocids', *Paleobiology*, vol. 23, no. 4, pp. 393–409, <www.psjournals.org/doi/abs/10.1666/0094-8373-23.4.393>.

10 Back from the dead

p. 151 The study of the shelf life of DNA was Morten E Allentoft et al. (2012), 'The half-life of DNA in bone: measuring decay kinetics in 158 dated fossils', *Proceedings of the Royal Society B*, vol. 279, no. 1748, pp. 4724–33, <rspb.royalsocietypublishing.org/content/early/2012/10/05/rspb.2012.1745.full>.

p. 152 Scott D Sampson's *Dinosaur Odyssey: Fossil threads in the web of life* was published in Berkeley by the University of California Press in 2009.

p. 154 The 2009 paper positing that *Dracorex hogwartsia* is actually a juvenile *Pachycephalosaurus* is John R Horner & Mark B Goodwin (2009), 'Extreme cranial ontogeny in the Upper Cretaceous dinosaur *Pachycephalosaurus*, *PLoS ONE*, vol. 4, no. 10, e7626, <www.plosone.org/article/info:doi/10.1371/journal.pone.0007626>.

p. 154 Mark Goodwin's comments to reporters appeared in Robert Sanders (2009), 'New analyses of dinosaur growth may wipe out one-third of species', media release, University of California, Berkeley, 30 October, <www.berkeley.edu/news/media/releases/2009/10/30_dino_demise.shtml>.

p. 156 Mark Ferguson's comments on hen's teeth were reported in (2006), 'Hens' teeth not so rare after all', media release, University of Manchester, 22 February, see <www.eurekalert.org/pub_releases/2006-02/uom-htn022206.php>.

p. 157 The quote from Jack Horner on 'building a dinosaur' come from his 2009 book with James Gorman, *How to Build a Dinosaur: Extinction doesn't have to be forever*, New York: Dutton.

11 The survival game

p. 163 The size and force of the dinosaur-destroying 'comet' is estimated by Walter Alvarez in his 1997 book *T. rex and the Crater of Doom*, Princeton, New Jersey: Princeton University Press.

p. 164 Walter Alvarez's description of the impact of the comet on the United States and Mexico also appears in his 1997 book *T. rex and the Crater of Doom*, Princeton, New Jersey: Princeton University Press.

p. 165 Richard Cowen's comments on the impact appear in his 2000 book *History of Life*, 3rd edn, Malden, Massachusetts: Blackwell Science.

p. 166 The study proposing a link between the solar system's passage through the Milky Way and the extinction of the dinosaurs is Dorminey, B (2008), 'Bouncing solar system killed the dinosaurs', *Cosmos*, 8 May, <www.cosmosmagazine.com/news/bouncing-solar-system-killed-dinosaurs>.

p. 167 The Alvarezes' paper on iridium levels is Luis W Alvarez et al. (1980), 'Extraterrestrial cause for the Cretaceous–Tertiary extinction', *Science*, vol. 208. no. 4448, pp. 1095–108, <www.sciencemag.org/content/208/4448/1095>.

p. 170 Discrepancies between dating methods are discussed in JE Fassett et al. (2011), 'Direct U–Pb dating of Cretaceous and Paleocene dinosaur bones, San Juan Basin, New Mexico', *Geology*, vol. 39, no. 2, pp. 159–62, <geology.gsapubs.org/content/39/2/159.abstract?ijkey=547f7b2ffdc038117ebbd8575a04a957e2db6af4&keytype2=tf_ipsecsha>.

p. 172 Some of Dave Hone's comments on bird survival appeared on his blog in 2013, 'How to survive mass extinction', 'Lost worlds', *The Guardian*, 20 September, <www.theguardian.com/science/lost-worlds/2012/sep/20/dinosaurs-fossils>.

p. 173 The 2009 study suggesting birds had the edge over dinosaurs due to their larger brains is Angela C Milner & Stig A Walsh (2009), 'Avian brain evolution: new data from Palaeogene birds (Lower Eocene) from England', *Zoological Journal of the Linnean Society*, vol. 155, no. 1, pp. 198–219, <onlinelibrary.wiley.com/doi/10.1111/j.1096-3642.2008.00443.x/full>.

References

An A–Z of feathered dinosaurs

The descriptions and/or announcements of the listed species appear in the papers noted below.

p. 177 *Anchiornis huxleyi*: Hu Dongyu et al. (2009), 'A pre-*Archaeopteryx* troodontid theropod from China with long feathers on the metatarsus', *Nature*, vol. 461, pp. 640–43, <www.nature.com/nature/journal/v461/n7264/full/nature08322.html>.

p. 177 *Anzu wyliei*: MC Lamanna et al. (2014), 'A new large-bodied oviraptorosaurian theropod dinosaur from the latest Cretaceous of western North America', *PLoS ONE*, vol. 9, no. 3, e92022, <www.plosone.org/article/info%3Adoi%2F10.1371%2Fjournal.pone.0092022>.

p. 178 *Archaeopteryx lithographica*: Angela Milner (n.d.), '*Archaeopteryx lithographica*', Natural History Museum, <www.nhm.ac.uk/nature-online/species-of-the-day/evolution/archaeopteryx-lithographica>.

p. 178 *Aurornis xui*: Pascal Godefroit et al. (2013), 'A Jurassic avialan dinosaur from China resolves the early phylogenetic history of birds', *Nature*, doi:10.1038/nature12168, <www.nature.com/nature/journal/vaop/ncurrent/full/nature12168.html>.

p. 178 *Avimimus portentosus*: Sergei M Kurzanov (1981), 'An unusual theropod from the Upper Cretaceous of Mongolia (fossil vertebrates of Mongolia)', *Transactions of the Joint Soviet–Mongolian Paleontological Expedition*, vol. 15, pp. 39–49.

p. 178 *Beipiaosaurus inexpectus*: Xu Xing et al. (1999), 'A therizinosauroid dinosaur with integumentary structures from China', *Nature*, vol. 399, pp. 350–54, <www.nature.com/nature/journal/v399/n6734/abs/399350a0.html>.

p. 179 *Caudipteryx zoui*: Ji Qiang (1998), 'Two feathered dinosaurs from northeastern China', *Nature*, vol. 393, pp. 753–61, <www.nature.com/nature/journal/v393/n6687/full/393753a0.html>.

p. 179 *Caudipteryx dongi*: Zhong-he Zhouh & Xiao-Lin Wang (2000), 'A new species of *Caudipteryx* from the Yixian Formation of Liaoning, Northeast China', *Vertebrata PalAsiatica*, vol. 38, no. 2, pp. 111–27, <www.ivpp.ac.cn/qt/papers/201206/P020120604544997033912.pdf>.

p. 179 *Citipati osmolskae*: James M Clark et al. (2001), 'Two new oviraptorids (Theropoda: Oviraptorosauria), Upper Cretaceous Djadokhta Formation, Ukhaa Tolgod, Mongolia', *Journal of Vertebrate Paleontology*, vol. 21, no. 2, pp. 209–13, <www.bioone.org/doi/full/10.1671/0272-4634%282001%29021%5B0209%3ATNOTOU%5D2.0.CO%3B2>.

p. 180 *Concavenator corcovatus*: Francisco Ortega et al. (2010), 'A bizarre, humped Carcharodontosauria (Theropoda) from the Lower Cretaceous of Spain', *Nature*, vol. 467, pp. 203–206, <www.nature.com/nature/journal/v467/n7312/full/nature09181.html>.

p. 180 *Conchoraptor gracilis*: Rinchen Barsbold (1986), 'Raubdinosaurier Oviraptoren', in El Vorobeva (ed.), *Gerpetologicheskie issledovaniya v Mongolskoi Narodnoi Respublike*, Moscow: Academy of Sciences, pp. 210–23.

p. 180 *Dilong paradoxus*: Xu Xing et al. (2004), 'Basal tyrannosauroids from China and evidence for protofeathers in tyrannosauroids', *Nature*, vol. 431, pp. 680–84, <www.nature.com/nature/journal/v431/n7009/full/nature02855.html>.

p. 180 *Eosinopteryx brevipenna*: Pascal Godefroit et al. (2012), 'Reduced plumage and flight ability of a new Jurassic paravian theropod from China', *Nature Communications*, vol. 4, article no. 1394, <www.nature.com/ncomms/journal/v4/n1/full/ncomms2389.html>.

p. 181 *Epidexipteryx hui*: Zhang Fucheng et al. (2008), 'A bizarre Jurassic maniraptoran from China with elongate ribbon-like feathers', *Nature*, vol. 455, pp. 1105–108, <www.nature.com/nature/journal/v455/n7216/full/nature07447.html>.

p. 181 *Gigantoraptor erlianensis*: Xu Xing (2007), 'A gigantic bird-like dinosaur from the Late Cretaceous of China', *Nature*, vol. 447, pp. 844–47, <www.nature.com/nature/journal/v447/n7146/full/nature05849.html>.

p. 182 *Jianchangosaurus yixianensis*: Hanyong Pu et al. (2013), 'An unusual basal therizinosaur dinosaur with an ornithischian dental arrangement from Northeastern China', *PLoS ONE*, vol. 8, no. 5, e63423, <www.plosone.org/article/info:doi/10.1371/journal.pone.0063423>.

p. 182 *Jinfengopteryx elegans*: Ji Qiang et al. (2005), 'First avialan bird from China (*Jinfengopteryx elegans* gen. et sp. nov.)', *Geological Bulletin of China*, vol. 24, no. 3, pp. 197–210, see <caod.oriprobe.com/articles/8775970/First_avialian_bird_from_China_Jinfengopteryx_elegans_gen__et_sp__nov_.htm>.

p. 182 *Juravenator starki*: Ursula B Göhlich & Luis M Chiappe (2006), 'A new carnivorous dinosaur from the Late Jurassic Solnhofen archipelago', *Nature*, vol. 440, pp. 329–32, <www.nature.com/nature/journal/v440/n7082/abs/nature04579.html>.

p. 182 *Microraptor zhaoianus*: Xu Xing et al. (2000), 'The smallest known non-avian theropod dinosaur', *Nature*, vol. 408, pp. 705–708, <www.nature.com/nature/journal/v408/n6813/abs/408705a0.html>.

p. 183 *Ningyuansaurus wangi*: Ji Qiang, Lü Jun-Chang, Wei Xue-Fang & Wang Xu-Ri (2012), 'A new oviraptorosaur from the Yixian Formation of Jianchang, Western Liaoning Province, China', *Geological Bulletin of China*, vol. 31, no. 12, pp. 2102–107, see <caod.oriprobe.com/articles/31641141/A_new_oviraptorosaur_from_the_Yixian_Formation_of_.htm>.

p. 183 *Nomingia gobiensis*: Rinchen Barsbold et al. (2000), 'A new oviraptorosaur (Dinosauria, Theropoda) from Mongolia: the first dinosaur with a pygostyle', *Acta Palaeontologica Polonica*, vol. 45, no. 2, pp. 97–106, <www.app.pan.pl/article/item/app45-097.html>.

p. 183 *Ornithomimus edmontonicus*: Darla K Zelenitsky et al. (2012), 'Feathered non-avian dinosaurs from North America provide insight into wing origins', *Science*, vol. 338, no. 6106, pp. 510–14, <www.sciencemag.org/content/338/6106/510>.

p. 184 *Oviraptor philoceratops*: Henry Fairfield Osborn (1924), 'Three new Theropoda, *Protoceratops* zone, central Mongolia', *American Museum Novitates*, no. 144, pp. 1–12, <digitallibrary.amnh.org/dspace/handle/2246/3223>.

p. 184 *Pedopenna daohugouensis*: Xu Xing & Zhang Fucheng (2005), 'A new maniraptoran dinosaur from China with long feathers on the metatarsus', *Naturwissenschaften*, vol. 92, no. 4, pp. 173–77, <link.springer.com/article/10.1007%2Fs00114-004-0604-y>.

p. 185 *Pelecanimimus polyodon*: Bernardino P Pérez-Moreno et al. (1994), 'A unique multitoothed ornithomimosaur dinosaur from the Lower Cretaceous of

Spain', *Nature*, vol. 370, pp. 363–67, <www.nature.com/nature/journal/v370/n6488/abs/370363a0.html>.

p. 185 *Protarchaeopteryx robusta*: Ji Qiang & Ji Shuan (1997), 'Protarchaeopterygid bird (*Protarchaeopteryx* gen. nov.) – fossil remains of archaeopterygids from China', *Geology in China*, vol. 238, no. 3, pp. 38–41, see <caod.oriprobe.com/issues/123938/toc.htm>.

p. 185 *Psittacosaurus mongoliensis*: Henry Fairfield Osborn (1923), 'Two Lower Cretaceous dinosaurs of Mongolia', *American Museum Novitates* no. 95, pp. 1–10, <http://digitallibrary.amnh.org/dspace/handle/2246/3267>.

p. 186 *Rahonavis ostromi*: Catherine A Forster et al. (1998), 'The theropod ancestry of birds: new evidence from the Late Cretaceous of Madagascar', *Science*, vol. 279, no. 5358, pp. 1915–19, <www.sciencemag.org/content/279/5358/1915>.

p. 186 *Scansoriopteryx heilmanni*: Stephen A Czerkas & Yuan Chongxi (2002), 'An arboreal maniraptoran from northeast China', in Stephen A Czerkas (ed.), *Feathered Dinosaurs and the Origin of Flight*, Blanding, Utah: The Dinosaur Museum, pp. 63–95, <www.dinosaur-museum.org/featheredinosaurs/arboreal_maniraptoran.pdf>.

p. 186 *Sciurumimus albersdoerferi*: Oliver WM Rauhut et al. (2012), 'Exceptionally preserved juvenile megalosauroid theropod dinosaur with filamentous integument from the Late Jurassic of Germany', *Proceedings of the National Academy of Sciences*, vol. 109, no. 29, pp. 11746–51, <www.pnas.org/content/109/29/11746>.

p. 187 *Shuvuuia deserti*: Luis M Chiappe et al. (1997), 'The skull of a relative of the stem-group bird *Mononykus*', *Nature*, vol. 392, pp. 275–78, <www.nature.com/nature/journal/v392/n6673/abs/392275a0.html>.

p. 187 *Similicaudipteryx yixianensis*: Tao He et al. (1998), 'A new genus and species of caudipterid dinosaur from the Lower Cretaceous Jiufotang Formation of Western Liaoning, China', *Vertebrata PalAsiatica*, vol. 46, no.3 , pp. 178–89, <www.ivpp.cas.cn/cbw/gjzdwxb/xbwzxz/200811/W020090813369309786831.pdf>.

p. 188 *Sinocalliopteryx gigas*: Ji Shuan et al. (2007), 'A new giant compsognathid dinosaur with long filamentous integuments from Lower Cretaceous of Northeastern China', *Acta Geologica Sinica*, vol. 81, no. 1, pp. 8–15, <www.geojournals.cn/dzxben/ch/reader/view_abstract.aspx?file_no=20070106&flag=1>.

p. 188 *Sinornithosaurus millenii*: Xu Xing et al. (1999), 'A dromaeosaurid dinosaur with a filamentous integument from the Yixian Formation of China', *Nature*, vol. 401, pp. 262–66, <www.nature.com/nature/journal/v401/n6750/abs/401262a0.html>.

p. 188 *Sinosauropteryx prima*: Chen Pei-Ji et al. (1998), 'An exceptionally well-preserved theropod dinosaur from the Yixian Formation of China', *Nature*, vol. 391, pp. 147–52, <www.nature.com/nature/journal/v391/n6663/full/391147a0.html>.

p. 189 *Tianyulong confuciusi*: Zheng Xiao-Ting et al. (2009), 'An Early Cretaceous heterodontosaurid dinosaur with filamentous integumentary structures', *Nature*, vol. 458, pp. 333–36, <www.nature.com/nature/journal/v458/n7236/full/nature07856.html>.

p. 189 *Velociraptor mongoliensis*: Henry Fairfield Osborn (1924), 'Three new Theropoda, *Protoceratops* zone, central Mongolia', *American Museum Novitates*, no. 144, pp. 1–12, <digitallibrary.amnh.org/dspace/handle/2246/3223>.

p. 190 *Xiaotingia zhengi*: Xu Xing et al. (2011), 'An *Archaeopteryx*-like theropod from China and the origin of Avialae', *Nature*, vol. 475, pp. 465–70, <www.nature.com/nature/journal/v475/n7357/full/nature10288.html>.

p. 190 *Yixianosaurus longimanus*: Xu Xing & Wang Xiao-Lin (2003), 'A new maniraptoran from the Early Cretaceous Yixian Formation of western Liaoning', *Vertebrata PalAsiatica*, vol. 41, no. 3, pp. 195–202, <www.ivpp.cas.cn/cbw/gjzdwxb/xbwzxz/200810/W020090813368479814753.pdf>.

p. 190 *Yutyrannus huali*: Xing Xu et al. (2012), 'A gigantic feathered dinosaur from the Lower Cretaceous of China', *Nature*, vol. 484, pp. 92–95, <www.nature.com/nature/journal/v484/n7392/full/nature10906.html>.

Image section

p. 5   Handy pose: Andrew RC Milner et al. (2009), 'Bird-like anatomy, posture, and behavior revealed by an Early Jurassic theropod dinosaur resting trace', PLoS ONE, vol. 4, no. 3, e4591, <www.plosone.org/article/info%3Adoi%2F10.1371%2Fjournal.pone.0004591>.

p. 6   Wonder wings (image on page 7): David WE Hone et al. (2010), 'The extent of the preserved feathers on the four-winged dinosaur *Microraptor gui* under ultraviolet light', *PLoS ONE*, vol. 5, no. 2, e9223, <www.plosone.org/article/info%3Adoi%2F10.1371%2Fjournal.pone.0009223>.

p. 9   Snack time: Lida Xing et al. (2012), 'Abdominal contents from two large early cretaceous compsognathids', *PLoS ONE*, vol. 7, no. 8, e44012, <www.plosone.org/article/info%3Adoi%2F10.1371%2Fjournal.pone.0044012>.

Glossary

amino acid – molecular building block of proteins.

bipedal – two-legged, rather than four-legged.

coelurosaurs – a group of theropod dinosaurs related to birds. It includes compsognathids, tyrannosaurs, ornithomimosaurs and maniraptorans.

Cretaceous – the geological period beginning 145 million years ago and ending 66 million years ago.

describe – in the context of discovery of new species the act of 'describing' or a 'description' means the formal task of naming and defining the characteristics of a novel plant or animal.

dinofuzz – fluffy down-like structures that represented an early stage of feather evolution.

dromaeosaurs – a group of small bird-like theropods (including *Velociraptor*) that were fast runners, effective predators and perhaps even pack hunters.

enantiornithes – Cretaceous-era offshoots of early birds that still had teeth.

formation – layers of rock laid down, without breaks to divide them, over a relatively short period of geological time.

genome – the entire sum of an organism's DNA.

Gondwana – southern supercontinent that began to break up into Africa, Australia, South America, India, Madagascar and Antarctica 184 million years ago.

hadrosaurs – the herbivorous duck-billed dinosaurs, a group of ornithischians that included *Maiasaura* and *Parasaurolophus*.

integumentary – structures originating from the skin of an organism, such as fluff, scales, fur and feathers.

Jurassic – the geological period beginning 201 million years ago and ending 145 million years ago.

keratin – the protein that makes up nails and claws, reptile scales, mammal fur and bird (and dinosaur) feathers.

maniraptors – theropod dinosaurs most tightly linked to birds. They include dromaeosaurs, oviraptorosaurs and therizinosaurs.

Mesozoic – the geological era comprising the Triassic, Jurassic and Cretaceous periods.

morphology – the form, shape and structure of an organism or physical feature.

ornithischian – one of the two divisions within the dinosaur group; the other is the saurischians. The ornithischian or 'bird-hipped' dinosaurs include long-necked sauropods and all the bipedal, predatory theropods.

oviraptorosaurs – a group of parrot-beaked omnivorous theropods with short pygostyle tails that are inferred to have had feather fans attached for display.

paedomorphosis – also known as neoteny, the retention of juvenile traits in an adult animal.

Palaeognathae – (Greek for 'old jaw') the most ancient branch of living birds, which includes tinamous and flightless ratites such as the kiwi, rhea, ostrich, cassowary and emu (and also the extinct elephant birds of

Madagascar and the moa of New Zealand). All other modern birds fall into the Neognathae group.

pennaceous – (pen-ay-SHUHS) a word used to describe feathers that are the typical modern feather shape, with a central shaft running the length and interlocking barbs running off to either side.

plumulaceous – (ploom-YUH-ley-SHUHS) a word used to describe feathers that have no central vane and are a messy jumble of filaments.

pneumatisation – invasion of air sacs into bird bones creating a lightweight honeycomb structure.

protofeather – simple filament representing one of the earliest stages of feather evolution.

pygostyle – the shortened tail structure of modern birds, to which a fan of feathers attaches.

quill knobs – pits where the ligaments of flight feathers attach to the arm bones of modern birds.

saurischian – one of the two divisions within the dinosaur group; the other is the ornithischians. The saurischian or 'lizard-hipped' dinosaurs, include heavy-set and armoured species and herd-living herbivores, such as the hadrosaurs.

soft tissue – any fossilised body part not created by the remains of bones ('hard tissue'), such as muscles, skin, internal organs, fur and feathers.

sternum – a bone between the ribs that acts like a keel and anchors the breast muscles that power the wings in modern birds.

stratigraphy – the study of the order and relative position of rock layers and how they relate to geological time.

taxonomy – biological discipline concerned with the classification and naming of organisms.

temporal paradox – the confusing problem in unravelling the evolution of flight whereby most of the feathered dinosaurs first discovered had lived significantly later than *Archaeopteryx* the 'first bird', and therefore could not have been ancestral to it.

terrestrial – land-living.

therizinosaurs – (ther-uh-ZIN-oh-SORE) carnivorous theropods whose teeth suggest they had returned to a vegetarian diet. The name means 'reaping' or 'scything' lizard.

theropods – a large group of bipedal saurischian dinosaurs that included all of the carnivorous species.

Triassic – the geological period beginning 252 million years ago and ending 201 million years ago.

troodontids – a group of small (less than 100-kilogram) theropods related to dromaeosaurs and birds. Examples include *Troodon*, *Anchiornis*, *Xiaotingia* and *Mei*.

warm-blooded – animals with high metabolic rates that can regulate their internal body temperature to within a narrow range, a process known as thermoregulation.

Pronunciation guide

Chicxulub – (CHIK-shoo-loob) a town on Mexico's Yucatán Peninsula.

Liaoning – (lee-ow-NING) a province in north-eastern China.

Select bibliography

Alvarez, W 1997, *T. rex and the Crater of Doom*, Princeton, New Jersey: Princeton University Press.

Benton, M 2014, *Vertebrate Palaentology*, 4th edn, Hoboken, New Jersey: Wiley-Blackwell.

Cowen, R 2000, *History of Life*, 3rd edn, Malden, Massachusetts: Blackwell Science.

Czerkas, SA (ed.) 2002, *Feathered Dinosaurs and the Origin of Flight*, Blanding, Utah: The Dinosaur Museum.

Foster, M & Lankester, E Ray (eds) 1902, *The Scientific Memoirs of Thomas Henry Huxley*, vol. 4, London: Macmillan.

Hanson, T 2012, *Feathers: The evolution of a natural miracle*, New York: Basic Books.

Horner, J & Gorman, J 2009, *How to Build a Dinosaur: Extinction doesn't have to be forever*, New York: Dutton.

Jaffe, M 2000, *The Gilded Dinosaur: The fossil war between E.D. Cope and O.C. Marsh and the rise of American science*, New York: Crown.

Lanham, U 2012, *The Bone Hunters: The heroic age of paleontology in the American West*, New York: Dover Publications.

Long, J 2012, *Dawn of the Deed: The prehistoric origins of sex*, Chicago: University of Chicago Press.

Long, J & Schouten, P 2008, *Feathered Dinosaurs: The origin of birds*, Melbourne: CSIRO Publishing.

Norell, M & Ellison, M 2005, *Unearthing the Dragon: The great feathered dinosaur discovery*, New York: Pi Press.

Paul, GS (ed.) 2000, *The Scientific American Book of Dinosaurs*, New York: Byron Preiss Visual Publications and Scientific American.

Sampson, SD 2009, *Dinosaur Odyssey: Fossil threads in the web of life*, Berkeley: University of California Press.

Shipman, P 1998, *Taking Wing: Archaeopteryx and the evolution of bird flight*, New York: Simon & Schuster.

Tudge, C 2009, *The Bird: A natural history of who birds are, where they came from, and how they live*, New York: Crown.

Wallace, DR 1999, *The Bonehunters' Revenge: Dinosaurs, greed, and the greatest scientific feud of the gilded age*, Boston: Houghton Mifflin.

Acknowledgments

Where to start with thanking those who made this book possible? I'd like to acknowledge all the experts who gave their time freely to chat to me, endured my pestering emails, sent me useful papers and suggestions, allowed me to use their artworks, answered numerous subsequent questions, and reviewed and commented on drafts of my chapters. These include: Paul Barrett, Mike Benton, Bhart-Anjan Bhullar, Luis Chiappe, Brian Choo, Julius Csotonyi, Phil Currie, Peter Dodson, Jeff Goertzen, Dave Hone, Jack Horner, John Long, Chris Organ, Richard Prum, Luis Rey, Alvaro Rosalen, Steve Salisbury, Peter Schouten, Paul Sereno, David Varricchio, Jakob Vinther, Lida Xing and Xu Xing.

I'd like to thank the British Museum in London and the State Library of New South Wales in Sydney for scouring their shelves to bring me tome after tome, and for providing places for quiet contemplation and study. I'm grateful to Jane McCredie and New-South Publishing for suggesting that I pitch this book in the first place; Yannick Lawry for motivating me to finally get the pitch together; and Australian science magazine *Cosmos*, which published my original feature on the evolution of birds from dinosaurs in 2010. I'd also like to thank Carolyn Reynolds for transcribing my interviews; Gary Nunn and Phil Currie for spending many, many hours looking over the text of every chapter and providing useful comments for revisions; and Elspeth Menzies, Emma Driver and Nicola Young for commenting on, editing and helping to improve the copy. Finally, I'm grateful to my mum for instilling in me a lifelong love of nature and natural history, and my dad for challenging me intellectually, inspiring me to work hard and – most especially – imploring me to study chemistry in my final years at school, without which the trajectory of my life would have been very different indeed.

Index

Index